Earth Systems Data and Models

Volume 4

Series editors

Bernd Blasius, Carl von Ossietzky University Oldenburg, Oldenburg, Germany
William Lahoz, NILU—Norwegian Institute for Air Research, Kjeller, Norway
Dimitri P. Solomatine, UNESCO—IHE Institute for Water Education, Delft, The Netherlands

Aims and Scope

The book series Earth Systems Data and Models publishes state-of-the-art research and technologies aimed at understanding processes and interactions in the earth system. A special emphasis is given to theory, methods, and tools used in earth, planetary and environmental sciences for: modeling, observation and analysis; data generation, assimilation and visualization; forecasting and simulation; and optimization. Topics in the series include but are not limited to: numerical, datadriven and agent-based modeling of the earth system; uncertainty analysis of models; geodynamic simulations, climate change, weather forecasting, hydroinformatics, and complex ecological models; model evaluation for decision-making processes and other earth science applications; and remote sensing and GIS technology.

The series publishes monographs, edited volumes and selected conference proceedings addressing an interdisciplinary audience, which not only includes geologists, hydrologists, meteorologists, chemists, biologists and ecologists but also physicists, engineers and applied mathematicians, as well as policy makers who use model outputs as the basis of decision-making processes.

More information about this series at http://www.springer.com/series/10525

Gloria Bordogna · Paola Carrara
Editors

Mobile Information Systems Leveraging Volunteered Geographic Information for Earth Observation

Springer

Editors
Gloria Bordogna
IREA CNR
Milano
Italy

Paola Carrara
IREA CNR
Milano
Italy

ISSN 2364-5830 ISSN 2364-5849 (electronic)
Earth Systems Data and Models
ISBN 978-3-319-89002-9 ISBN 978-3-319-70878-2 (eBook)
https://doi.org/10.1007/978-3-319-70878-2

Printed on acid-free paper

This Springer imprint is published by Springer Nature
The registered company is Springer International Publishing AG
The registered company address is: Gewerbestrasse 11, 6330 Cham, Switzerland

Contents

Introduction

Global Earth observation demands both ubiquitous worldwide interaction and ground-based information. The widespread diffusion of the Web and smart devices allows everybody to access and create georeferenced distributed multimedia information, termed volunteered geographic information (VGI) after Goodchild [1]. By the simple use of applications (apps) or websites accessing mobile information systems, everybody can provide updated multimedia content from everywhere. Smart devices are generally equipped with both low-cost memories and sensors such as GNSS (global navigation satellite system) receivers using the US global positioning system (GPS) for location awareness, and cameras and accelerometers for creating and geotagging pictures and videos, i.e., to associate them with a geographic footprint on Earth. Besides images and videos, VGI often includes free text describing observations by the author, categories of classification of the observed entities, recordings of sounds, which provide rich and both subjective and objective descriptions of reality.

Volunteered geographic information is being investigated as a novel opportunity to enrich research projects with widespread ground data that can be exploited in Earth observation, including monitoring of natural, environmental, human-driven and social changes and events. In these contexts, VGI can be considered a relevant aspect of citizen science (CS), i.e., science made by citizens' contributions. For this reason, in the last years an increasing number of academic papers and special journal issues have focused on the characteristics of VGI within CS projects and on evaluations of the potential for VGI to help scientists, policymakers, and business companies in conceiving and launching new services.

Volunteered geographic information about Earth observation could be fruitfully exploited to increase awareness and knowledge of people located in a territory. Preparedness and emergency management, tourism, leisure and entertainment recommendations, social security and many scientific areas, such as climatology, ecology, biodiversity, agriculture and marine research are only some of the fields that might benefit from the use of mobile information systems leveraging VGI. These diverse possible uses of VGI motivate the attention of both business companies and industry, willing to offer low-cost location-based services to their customers, and research communities, operational organizations and NGO in need of huge amounts of widespread in situ georeferenced data to carry out their studies, data that they could hardly acquire otherwise, especially in the cases of low-budget projects.

Though the areas of mobile-based information and VGI are related, until now there has not been a comprehensive analysis of their connections and synergies. On the one side, the research literature on mobile information systems has a number of dedicated conferences and journals [2–4]. Special journal issues have dealt with mobile information systems and computing but have not considered VGI [5]. Only very recently special issues about the management of crowdsourced geospatial information in the context of mobile information systems have been published, and there are open calls for papers at the date of writing this book [6–9].

On the other side, in the last years, VGI and crowdsourced geo-information have become topics of interest in many journals on geo-information and conferences related to databases, geographic information systems and information retrieval such as ACM CIKMs, ACM GIS, AGILE, ACM SIGIR, IEEE/ACM/WIC WI, IAR Track of ACM SAC. Dedicated workshops on several aspects of VGI management, such as the recent AGILE 2017 Workshop on VGI analytics, have been organized [10]. Furthermore, there are a number of special issues of journals which are related to several aspects of VGI management and use [12–16], but they do not cover specific aspects related to the acquisition of VGI by mobile information systems for Earth observation applications.

With respect to the published literature, the contents of the present volume stress the link between VGI creation by smart devices and its use in Earth observation, in particular within both experimental and basic research perspectives. Specifically, we want to offer some possible answers to open questions such as:

- *Can experiences carried out within VGI CS projects (collecting in situ georeferenced information for Earth observation) provide useful insights on the characteristics of volunteers and their influence on created VGI?* This is an important question whose answers can help identify “a priori” criteria for selecting volunteers depending on the tasks requested within a CS project.
- *What are the characteristics of user interfaces for VGI creation on smart devices that can help engage volunteers to create high-quality VGI?* Analysing distinct application needs of Earth observations projects (which demand categorization of observations based on the perceived or measured properties of the objects/events which are targeted by VGI) may suggest both adequate means and external knowledge for supporting the creation of high-quality VGI.
- *From lessons learnt within Earth observation projects can we suggest directions for improving VGI reuse by means of novel components and/or web services of mobile information systems? That is, can we suggest methods for enabling VGI stakeholders to select relevant VGI for their specific tasks and needs?* This is an important issue in the data value chain, since the value of data greatly depends on the potential of its use and reuse. Providing means to aid stakeholders

selecting VGI (which was created within other projects) for some specific task of their interest can optimize the chain from data creation to data (re)use.

- *Which geo-information algorithms and tools can help VGI management for Earth observation within a mobile information system context? That is, how can we fruitfully use VGI acquired by smart devices to cross-analyse and fuse it with multisource information?* This is a crucial task in the age of big geo data, characterized by huge volumes, high velocity and variety, questionable validity and veracity. More frequently, we will have to cope with redundant co-referent VGI, created within distinct applications by distinct means (both employing distinct sensors and software) relative to the observation of the same events/entities of the real world. Let us think of the many smart applications that allow volunteers to point out traffic jams. In order to increase VGI quality, one can exploit the fusion of information from distinct sources by filtering and weighing the contributions. To this end, new methods are needed to identify the highest quality sources of VGI for a given purpose and to quantify and compare their quality. In this respect, the characteristics of the sources and the VGI lineage will become essential for the definition of VGI metadata.

Discussions on the above questions, which beg some possible answers, solutions and open issues, are dealt within the chapters; they are organized into three parts, introducing the reader to three basic aspects involved in VGI creation by mobile information systems and its use for Earth observation.

The first part focuses on the analysis of volunteers' engagement in relation to their expertise, motivation, types the task they are called to perform, means and tools provided to them for creating VGI. Constant volunteers' engagement over time is essential to the success of any collaborative initiative related to Earth observation and monitoring. Nevertheless, it is often difficult attracting volunteers to contribute within a project and it is even more difficult to maintain their constant engagement. The three chapters in this part describe several experiences of VGI collection within CS initiatives in relation to the analysis of the volunteers' characteristics and outline of the lessons learnt. Specifically, the first chapter describes several collaborative outdoor and indoor mapping experiments using either VGI created by "traditional" mapping tools developed for the popular Open Street Map project, emotional and game-based mapping methods or crowd-contributed geographic contents created implicitly within *Twitter*. The second chapter discusses the experience of the *Crowd4Sat* project, funded by the European Space Agency, that investigated different facets of how crowdsourcing and citizen science impact upon the validation, use and enhancement of products and services exploiting observations from satellites. The third chapter focuses on an experiment of VGI apps used to collect either biological or abiotic observations during an excursion following naturalistic trails of the Italian Alps connecting sites of the Long Term Ecological Research (LTER) network; it analyses the impact on the practices of the research communities when they accept to use VGI created by means of innovative smart apps.

The second part of the volume is dedicated to the management of VGI within information systems for Earth observation, i.e., to introduce methods, techniques and algorithms that can be applied to exploit VGI for distinct Earth observation purposes. The first chapter in this part proposes the use of VGI for creating "visual ontologies" for Earth observation, which could be exploited to design novel smart applications for mobile information systems aiding volunteers to easily create VGI. To generate visual ontologies, the chapter proposes to take advantage of a lot of pictures provided by volunteers for both creating and organizing the domains of visual concepts to be used in scientific projects, for example, for the purpose of monitoring glaciers. The second chapter evaluates the performance of commercially available, low-cost sensors integrated within a mobile information system for air quality mapping and communicating through the mobile network or directly linking to smartphones carried by users. It demonstrates that by exploiting the "swarm knowledge" of the entire network of sensors through the adoption of geostatistical data fusion techniques, useful information from the data can be extracted even though individual sensors are subject to significant uncertainty. The third chapter of the part, by discussing an experience of monitoring urban mobility based on the use of the *Moves* smart app developed by *Facebook*, motivates how novel data science methods are needed for an effective VGI management and cross-analysis with multisource information such as open data, VGI from distinct projects, and authoritative geo-information.

The last part of the volume focuses on the crucial aspects of VGI questionable quality and the proposal of methods and techniques to cope with it. One of the main objections of VGI detractors is that VGI is invariably affected by incompleteness, inaccuracy, imprecision and uncertainty, and thus one cannot trust and use it. The first chapter in this part analyses the quality challenges that projects using VGI have to face. The second chapter stresses the importance of creating and maintaining good metadata for VGI to improve its quality and to ensure the appropriate combination and reuse of the resulting datasets. In this respect, authors discuss major challenges and present a set of examples of current practices. The last two chapters propose VGI quality assessment methods on the stakeholder side: the first of them is inspired by a decision-making model that can handle the distinct reputations of the information sources and take into account multiple evaluation criteria. The second method adopts fuzzy ontologies and level-based reasoning in a database framework to represent VGI affected by imprecision and uncertainty to control and filter information from volunteers based on consumers' quality needs.

Acknowledgements The chapters have been written by well-known researchers and academic and research groups internationally active in the field of VGI and CS. We want to express our gratitude to the chapter authors, to the anonymous referees whose excellent work helped us to improve the contents, to the series editors for their comments and suggestions, and finally Springer Verlag for the assistance in producing this publication.

Gloria Bordogna
Paola Carrara

References

1. Goodchild, M.F., (2007). Citizens as voluntary sensors: spatial data infrastructure in the world of web 2.0. *International Journal of Spatial Data Infrastructures Research*, 2, 24–32.
2. MobiWIS—International Conference on Mobile Web and Information Systems, since 2013. https://link.springer.com/conference/mobiwis. Accessed July 4, 2017.
3. MOBIS—IFIP Working Conference on Mobile Information Systems, 2004 and 2005, http://dl.ifip.org/db/conf/mobis/index.html. Accessed July 4, 2017.
4. Hindawi open access Journal on Mobile Information Systems, https://www.hindawi.com/journals/misy/. ISSN: 1574-017X (Print), ISSN: 1875-905X (Online), doi:10.1155/9071. Accessed July 4, 2017.
5. Middleton, C.A., Scheepers, R., Tuunainen, V., (2011). Special issue on mobile information systems and mobility. *European Journal of Information Systems*. (WWW document), http://www.palgrave-journals.com/ejis/index.html. Accessed October 1, 2011.
6. Chen, L., Julien, O., Lohan, E.S., Seco-Granados, G., Chen, R. (2017). Special issue on Mobile Geospatial Computing Systems for Ubiquitous Positioning, in open access journal *Mobile Information Systems*, Hindawi, https://www.hindawi.com/journals/misy/si/343859/. Accessed July 4, 2017.
7. Yim, J., Ganesan, S., Kang, B.H. (2017). Location-Based Mobile Marketing Innovations, in open access journal *Mobile Information Systems*, Hindawi, https://www.hindawi.com/journals/misy/si/763134/. Accessed July 4, 2017.
8. Calafate C.T., Wu C., Natalizio E., Martínez, F.J. (2016). Crowdsensing and Vehicle-Based Sensing, in open access journal *Mobile Information Systems*, Hindawi, https://www.hindawi.com/journals/misy/si/154290/. Accessed July 4, 2017.
9. Huerta J., Gould M., Dos Santos V.M.P.D., Torres-Sospedra T. (2018). Call for papers for the special issue Geospatial Mobile Solutions for Smart Cities, in open access journal *Mobile Information Systems*, Hindawi, https://www.hindawi.com/journals/misy/si/849837/cfp/. Accessed July 4, 2017.
10. Pfoser, D., Voisard, A. (2013). GEOCROWD workshop report: The second int'l workshop on crowdsourced and volunteered geographic information 2013: (Orlando, FL—Nov. 5, 2013). SIGSPATIAL Special (vol. 6, iss. 1, p. 11, 2014).
11. Mooney, P., Zipf, A., Jokar, J., Hochmair, H.H. (2017). AGILE Workshop on VGI-Analytics, http://www.cs.nuim.ie/~pmooney/VGI-Analytics2017/. Accessed July 4, 2017.
12. Mooney, P., Zipf, A., Jokar, J., Hochmair, H.H. (2017). Special issue on Volunteered Geographic Information (VGI)-Analytics, forthcoming in Geo-spatial Information Science, late 2017. http://explore.tandfonline.com/cfp/est/gsis/si3. Accessed July 4, 2017.
13. Goodchild, M., Aubrecht, C., Bhaduri, B. (2016). Special issue Role of Volunteered Geographic Information in Advancing Science, Transactions in GIS, http://onlinelibrary.wiley.com/doi/10.1111/tgis.12242/full. Accessed July 4, 2017.
14. Zipf, A., Resch, B. (2015). Special issue on GeoWeb 2.0, *ISPRS International Journal of Geo-Information*, http://www.mdpi.com/journal/ijgi/special_issues/geoweb-2.0. Accessed July 4, 2017.
15. See, L, Fritz, S., de Leeuw, J., (2013). Special Issue Collaborative Mapping, *ISPRS International Journal of Geo-Information*, 2(4), 955–958. doi:10.3390/ijgi2040955.
16. Zipf, A., Jonietz, D., Antoniou, V., See, L. (2017). Special Issue Volunteered Geographic Information, *ISPRS International Journal of Geo-Information*, http://www.mdpi.com/journal/ijgi/special_issues/VGI. Accessed July 4, 2017.

Part I
Experiences of VGI Creation & Exploitation for Citizen Science Projects

Mapping Parties at FOSS4G Europe: Fun, Outcomes and Lessons Learned

Maria Antonia Brovelli, Peter Mooney, Ludovico Biagi, Marco Brambilla, Irene Celino, Eleonora Ciceri, Nicola Dorigatti, Haosheng Huang, Marco Minghini and Vijaycharan Venkatachalam

Abstract Since OpenStreetMap (OSM) appeared more than ten years ago, new collaborative mapping approaches have emerged in different areas and have become important components of localised information and services based on localisation. There is now increased awareness of the importance of the space-time attributes of almost every event and phenomenon. Citizens now have endless possibilities to quickly geographically locate themselves with an accuracy previously thought impossible. Based on these societal drivers, we proposed a number of collaborative mapping experiments ("mapping parties") to delegates of a large open-source geospatial conference and to citizens of the conference's host city during July 2015.

M. A. Brovelli (✉) · L. Biagi · M. Minghini
Department of Civil and Environmental Engineering, Politecnico di Milano, Piazza Leonardo da Vinci 32, 20133 Milan, Italy
e-mail: maria.brovelli@polimi.it

L. Biagi
e-mail: ludovico.biagi@polimi.it

M. Minghini
e-mail: marco.minghini@polimi.it

P. Mooney
Department of Computer Science, Maynooth University, Eolas Building, Maynooth, Co. Kildare, Ireland
e-mail: Peter.Mooney@nuim.ie

M. Brambilla
Dipartimento di Elettronica, Informazione e Bioingegneria (DEIB), Politecnico di Milano, via Ponzio 34/5, 20133 Milan, Italy
e-mail: marco.brambilla@polimi.it

I. Celino
CEFRIEL, via Fucini 2, Milan, Italy
e-mail: irene.celino@cefriel.com

E. Ciceri
Dipartimento di Elettronica, Informazione e Bioingegneria (DEIB), Politecnico di Milano, Como Campus, via Ponzio 34/5, Milan, Italy
e-mail: eleonora.ciceri@polimi.it

G. Bordogna and P. Carrara (eds.), *Mobile Information Systems Leveraging Volunteered Geographic Information for Earth Observation*, Earth Systems Data and Models 4, https://doi.org/10.1007/978-3-319-70878-2_1

These mapping parties had a wide conceptual range from VGI (Volunteered Geographic Information) to geo-crowdsourcing (involuntary crowd-contributed geographic content). Specifically, the mapping parties were: (1) "traditional" OSM mapping, (2) indoor mapping and localisation, (3) emotional mapping of cities, (4) game-based validation of land cover data sets and (5) sensing the city and conference implicitly from Twitter. In this chapter, we outline the aims, structure and implementation of these experiments. We discuss the key outcomes and lessons learned from each of the mapping experiments in order to demonstrate the commonalities and also the differences between experiments. We consider future research directions for collaborative mapping approaches.

1 Introduction

In July 2015, Politecnico di Milano, Como Campus, organised the FOSS4G (Free and Open Source Software for Geomatics) Europe conference (http://europe.foss4g.org/2015). This conference is held biannually when the global FOSS4G conference is held outside Europe which was the case in 2015 when FOSS4G was held in South Korea during September 2015. This conference brings together a very wide spectrum of delegates and attendees including FOSS4G software developers, users, educators, researchers, industry users of FOSS4G, open-source software and open data advocates. As organisers of the FOSS4G conference, we decided that it would be an interesting and challenging idea to organise mapping parties within the conference. One of the struggles faced by mapping parties is attracting and motivating people to participate. With almost 400 delegates attending the conference the majority of which had experience in mapping, GIS, geographic data collection, we felt that this provided an excellent opportunity to organise mapping parties which would attract highly skilled participants from a diverse set of backgrounds and disciplines. We decided to organise the mapping parties as a voluntary activity for delegates during free time at the conference. Specifically, the mapping parties were organised and timetabled between the final session of the first day of the conference and the conference icebreaker party (http://europe.foss4g.org/2015/Mapping%20parties). The time between these two events was approximately three hours.

N. Dorigatti
Trilogis Srl, Via Zeni 8, 38068 Rovereto (Trento), Italy
e-mail: nicola.dorigatti@trilogis.it

H. Huang
Department of Geodesy and Geoinformation, Vienna University of Technology, Karlsplatz 13, 1040 Vienna, Austria
e-mail: haosheng.huang@geo.uzh.ch

V. Venkatachalam
GESP Srl, Viale Scarampo 47, 20148 Milan, Italy
e-mail: vijaycharan@gesp.it

Specifically, the mapping parties were: (1) a traditional OpenStreetMap mapping party, (2) indoor mapping and localisation, (3) emotional mapping of cities, (4) game-based validation of land cover data sets and (5) sensing the city and conference implicitly from Twitter. During the opening welcome session of the conference, all mapping party organisers were given the opportunity to advertise their mapping party and call for participation from delegates. Each mapping party organiser presented their mapping party to the audience. Information was communicated over social media and using the official conference website.

In this chapter, we outline the aims, structure and implementation of these mapping parties. We discuss the key outcomes and lessons learned from each of the mapping parties in order to demonstrate the commonalities and also the differences between experiments. We consider future research directions for collaborative mapping approaches based on our experiences at FOSS4G Europe 2015. This chapter is organised as a series of descriptions of the individual mapping parties. Each of these descriptions follows a specific template. Each mapping party is described with a short introduction (Organisation and Set-up). The mapping party organisers are then asked to describe: the strategy for data collection in the mapping party (Strategy), the participation levels in their mapping party (Participation), results arising from the mapping party (Results and Outcomes) and an overall summary of the key lessons learned (Lessons and Evaluation). The paper closes with the conclusion section, where we discuss the aspects of fun, mapping and overall lessons learned from this experience at FOSS4G Europe 2015.

2 OpenStreetMap Mapping Party (MP1)

2.1 MP1: Organisation and Set-up

The first mapping party held during FOSS4G Europe 2015 was a traditional OpenStreetMap (OSM) mapping party. Founded in 2004, OSM is nowadays the most popular project of Volunteered Geographic Information (VGI; [1]).

Its purpose is to create a free, editable map of the world built by volunteers and released with an open content license (http://wiki.openstreetmap.org/wiki/About_OpenStreetMap). At the time of writing (January 2017), there are more than three million registered contributors to OSM (http://wiki.openstreetmap.org/wiki/Stats). The literature has shown that although the greatest percentage of the edits is performed by relatively few users (see, e.g., [2]), the OSM database is constantly growing at a very fast rate. There are mainly three ways users can contribute data to OSM: by digitising map objects visible on aerial or satellite imagery (armchair mapping); by surveying data on the field (field mapping); and by importing open-licensed data sets (bulk import). A typical application of digitisation from aerial or satellite imagery is for humanitarian purposes, e.g. after disaster strikes [3]. Bulk import is usually performed by experts as it is a very tricky and potentially dangerous operation. The third mapping strategy—field mapping—was the one proposed for the OSM mapping party at FOSS4G Europe 2015. This kind of experience has a long and successful tradition within the OSM community. Most of the activities advertised on the OSM events wiki page (http://wiki.openstreetmap.org/wiki/Current_events) are field mapping events or “mapping parties”. The main characteristics of these mapping parties along with their organisational, social and societal aspects/implications are also discussed in the literature (see, e.g., [4–6]).

2.2 MP1: Strategy

The mapping party was conceived and initiated by a small number of active OSM contributors participating to the conference, including one living in Como who had the best local knowledge of the city structure. It was decided to focus the mapping efforts on Points of Interest (POIs) which were largely missing in Como city, including commercial activities (shops, hotels, restaurants, etc.) and other common objects such as building numbers and trash cans. The OSM mapping party was opened to both people joining FOSS4G Europe 2015 and external people such as local OSM contributors. For this reason, the mapping party was advertised not only on the conference website and social medias, but also on the official OSM events page (http://wiki.openstreetmap.org/wiki/Past_Events_2015) through a dedicated wiki (http://wiki.openstreetmap.org/wiki/Como_foss4ge_2015). To allow data collection more equally distributed over Como city and avoid the duplication of efforts, the basic idea was to subdivide the participants into groups, each of which was lead by an expert OSM contributor. Como city centre was divided into five areas having almost the same extent and located between the conference venue, where the mapping party began, and the icebreaker party venue, where the mapping party ended (see Fig. 1). The strategy adopted to collect OSM data on the field was based on Field Papers (http://fieldpapers.org). These field papers consist of OSM maps which can be printed, used on the field to draw objects and write down notes and finally digitised (either in a geo-referenced scanned version or in the original paper version) to add OSM data. Field papers were chosen for their ease of use which made them suitable for people new to OSM considering joining the mapping party.

2.3 MP1: Participation

The OSM mapping party was among the most well attended. Despite the very hot weather conditions, 40 people participated who were unevenly distributed into 16 men and 24 women. After participants were instructed on the kind of objects to survey and how to use field papers, five groups of eight people were formed according to personal choices as well as linguistic constraints. In turn, most of the groups further split into subgroups of 2–3 people in order to subdivide the work and

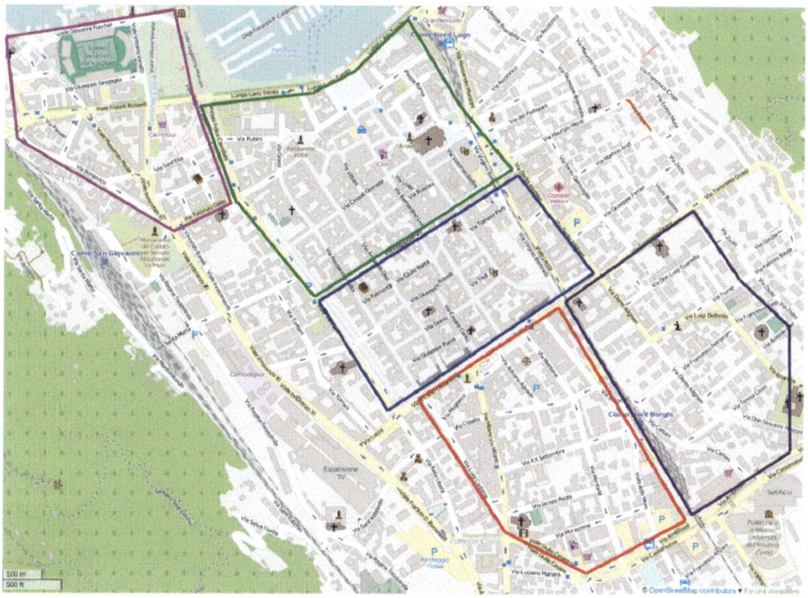

Fig. 1 Five areas of Como city centre where the OSM mapping party took place

survey their area in the shortest possible time. One of the field papers annotated during the mapping party is shown in Fig. 2.

On the following day, a data upload session was scheduled during the conference lunch break with the purpose of showing participants how to add the surveyed data to the OSM database. A brief tutorial was given on how to upload data and assign the correct tags using the two most used OSM editors, namely iD (http://wiki.openstreetmap.org/wiki/ID) and JOSM (https://josm.openstreetmap.de). Due to the need to first train participants in using the software as well as the short duration of the session (one hour), a very small portion of the surveyed data was uploaded during this session. In addition, far less than the 40 people who had joined the mapping party attended the data upload session. Many of them had already left their field papers to the organisers just after the mapping party, while many others did the same after the data upload session. This was mainly due to the participants' lack of time or lack of confidence in autonomously uploading data. Only some of the participants kept their field papers and promised to upload the data by themselves at a later stage. As a result, the mapping party organisers were given almost 2/3 of field papers and they worked hard to upload data in OSM during the remaining two conference days.

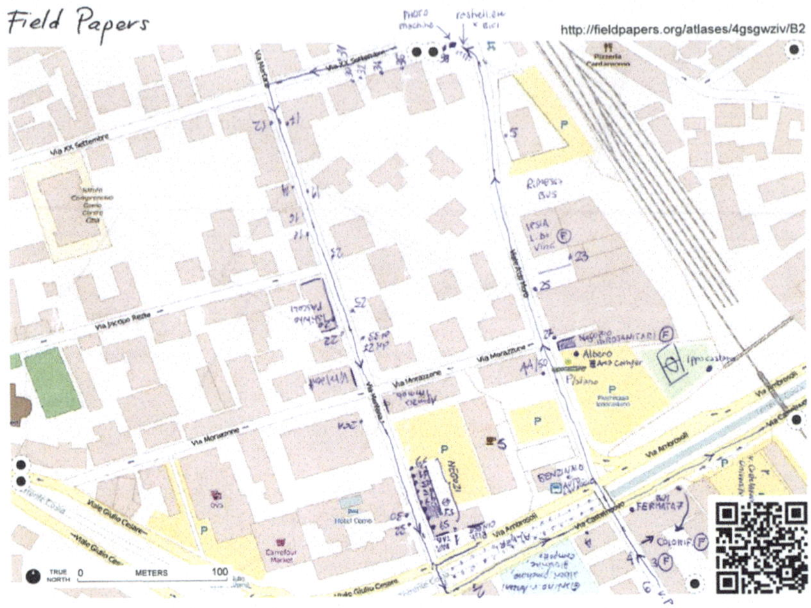

Fig. 2 A field paper used during the mapping party showing POIs annotations

2.4 *MP1: Results and Outcomes*

By the last day of the conference, a considerable amount of new POIs became available in the OSM map of Como city as a result of the mapping party. However, the upload of the data collected during the mapping party continued over the following weeks. To quantify the mapping effort, the number of new nodes added to the OSM map of Como city during the first week after the mapping party was about 700. This has brought a considerable enrichment to the map and demonstrated the high success of the event. Still most of the POIs visible on the current (January 2017) OSM map of Como city centre (http://osm.org/go/0CkuNw ~ HQ-) were added during the mapping party in July 2015 (see Fig. 3).

2.5 *MP1: Lessons Learned and Evaluation*

There are a number of lessons learned from this experience. Both the mapping party and the data upload session happened in an inclusive, relaxed and friendly atmosphere which was in line with the general atmosphere of FOSS4G conferences, where most of the people are young and community building plays a primary role. We maintained the social dimension of mapping parties which is also crucial, as highlighted by Perkins and Dodge [5]. We have no hesitation in recommending

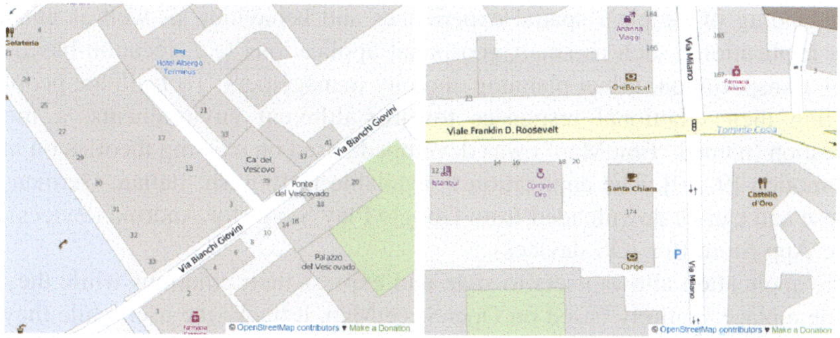

Fig. 3 Portions of the OSM map of Como city centre showing POIs (mainly shops and building numbers) added during the mapping party

OSM mapping parties as side events at future FOSS4G conferences. These mapping parties can benefit from the fact that there are typically many OSM experienced users within the same FOSS4G community. Some observations are also worth mentioning which will support better organisation of similar mapping events in the future. Despite being very easy and intuitive tools, field papers also showed some drawbacks. First, by their own nature, field papers limit the range of geospatial information which is actually recordable [7]. In addition, the amount and density of the annotations and drawings on field papers often required the very same person who used the field paper to perform data upload. As mentioned above, this was not always possible because a large number of participants left their field papers to the organisers. In contrast to other mapping parties [5], the data upload session was too short and allowed the upload of only a small part of the information surveyed. Many participants did not have time to become sufficiently familiar with the OSM editors to feel that they can autonomously upload data after the event. However, this time limitation was very much influenced and constrained by the FOSS4G conference schedule.

3 Emotional Mapping (MP2)

3.1 MP2: Organisation and Set-up

The emotional mapping party, which is extended from the EmoMap project at TU Wien in Austria [8], aims to collect people's emotional responses towards environments when travelling in an environment such as the city of Como. Research on environmental psychology suggests that all stimuli, including environments of different scales, are perceived not only according to physical features, but also affectively—in terms of a person's emotional responses to them. For example, some places are experienced as unsafe or boring, while some others are experienced as attractive and interesting. Studying these emotional responses contributes to a better understanding of people's spatial experiences and behaviour, as well as enabling many applications, such as smart geospatial applications (e.g. location-based services), transportation, urban planning and city management. To facilitate people in reporting their emotional responses towards different environments, a mobile application (named "EmoMap") was developed based on existing theories on affect and emotion [9, 10]. The application is available in English, Italian, German and Chinese and can be downloaded from Google Play Store (for Android devices) and Apple App Store (for iOS devices).

The application allows users to state and express their emotions while they are visiting a place. Entirely based on OpenStreetMap, it tracks the user while they are navigating a place of interest, always pointing at their current position, so that a user can state how they feel about their current surroundings. Together with the emotions they feel, the user can provide a subjective brief description of the place by specifying a set of adjectives (e.g. active, boring, disgusting, frightening) and some information about his visit (i.e. the frequency with which they visit the place and their current company). This last piece of information is needed to understand if the emotion the user specifies is somehow shifted towards a positive or negative flavour according to the familiarity with the place and the pleasure of being with other people.

3.2 MP2: Strategy

Generally, whenever an Internet connection (either via Wi-fi or with a mobile connection) is available, the user is able to tag any place in the world with an emotion. However, this mapping party could involve some people without an Internet connection. Thus, to run properly during the emotional mapping party, an offline version of the application has been provided. Such version was provided with the map of Como city already included in the downloaded application, so that it was possible to use this map without downloading it in an online fashion. In order to help users locate themselves, this application shows an OpenStreetMap map with a marker in the centre indicating the current location. Users can also adjust their current location if the GPS location does not match their surroundings. Users can then adjust the slider to indicate their "level of comfort" at the current environment, on a seven-point Likert scale [11], from uncomfortable ("1") to neutral ("4") and comfortable ("7"). And then, users are asked to select a subset of adjectives from a list of environment-related emotional adjectives. This list was introduced by Russell and Pratt [12] to describe the affective quality of places. In the next step, users are asked to provide some contextual information about their visit, particularly on company ("with whom?") and familiarity with the current place ("first time here?"). Figure 4 provides some screenshots of these options.

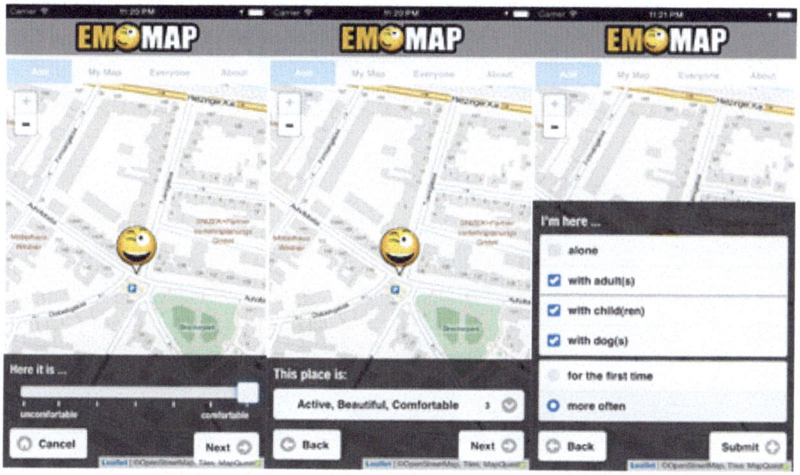

Fig. 4 Screenshots of the mobile application for collecting people's affective responses to environments (map data: © OpenStreetMap contributors, CC-BY-SA). The marker in each map centre denotes the current location. Users can also move the marker to adjust their location. Please note that we explicitly ask users to focus on the environment ("Here it is …")

3.3 *MP2: Participation*

The emotional mapping party was promoted to the conference participants as well as to the general public in Como via local newspaper "La Provincia" and local TV "Espansione". To motivate participation, we provided awards and gifts to the top three people who contributed the most overall. In total, 94 people participated in the mapping party and they contributed 244 emotional responses. The average age of the contributors was 38.7 years with a standard deviation of 11.6. On average, the number of contributions per user was 2.59 with a standard deviation of 6.30. The general contribution patterns follow a long tail distribution. In fact, only 56 (out of 94) people actually contributed one or more emotional responses. Among these 56 people, only seven of them contributed more than five responses.

3.4 *MP2: Results and Outcomes*

We analysed these emotional responses to understand how people perceived and evaluated the Como city. Figure 5 shows an overview of all the emotional responses within Como city. In general, emotional responses are mostly positive. The mean value of these "level of comfort" ratings is 5.12 with a standard deviation of 1.65, on a scale of 1 ("very uncomfortable") to 7 ("very comfortable"). Contributors mostly perceived the study area as "Enjoyable", "Beautiful", "Comfortable" and "Active". To have a closer look at the results, we visualised

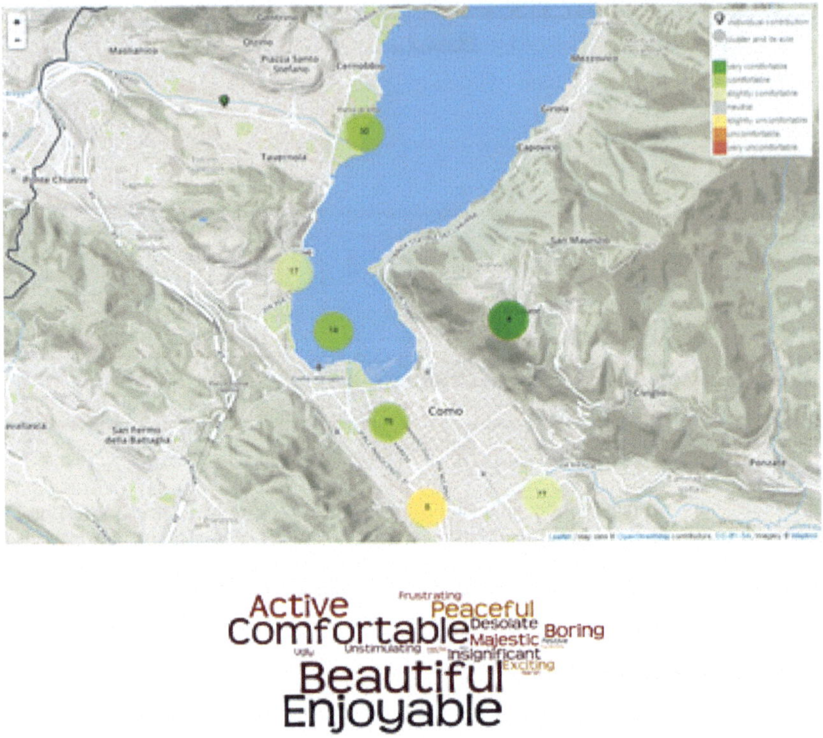

Fig. 5 Overview of the emotional responses within the Como city. Top: "level of comfort" ratings: colours of the markers indicate values of the "level of comfort" ratings, with green being comfortable, grey being neutral and red/yellow being uncomfortable. Bottom: tag cloud of the environment-related emotional adjectives

each of these affect responses as a single marker (Fig. 6). As shown in Fig. 6, affective responses around the Como Lake and the city centre of Como were very positive, while there are several "negative" spots, especially around "Mercato Comunale" ("Municipal Market"), "Como Nord Borghi" and "Via Achille Grandi".

3.5 *MP2: Lessons Learned and Evaluation*

The outcome of this mapping party confirms our expectation that different environments were interpreted differently and consequently caused different emotional responses. Therefore, as a future work, it might be interesting to correlate people's emotional responses and the environmental characteristics of different places. To address this issue, more emotional data should be collected, as well as techniques on assuring and assessing the data quality should be developed. Furthermore, due to the complex nature of emotion, it is also important to collect more contextual information

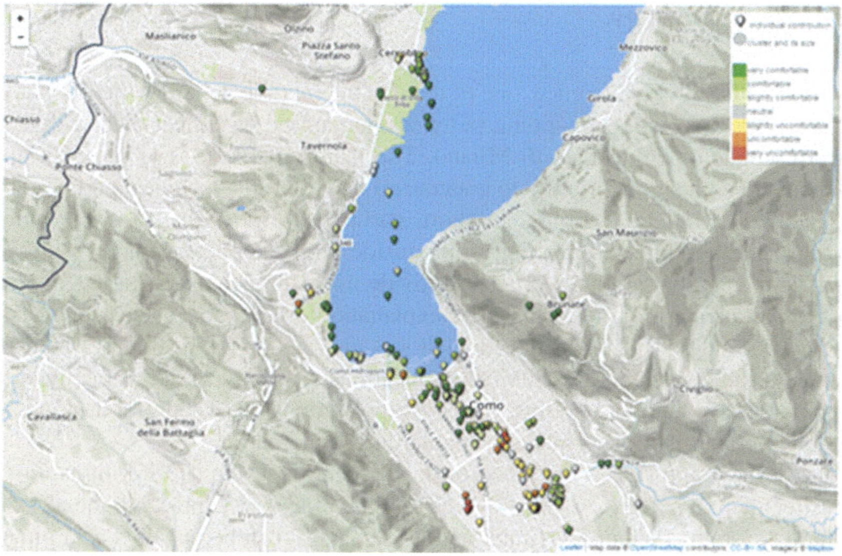

Fig. 6 Emotional responses within the study area. Each response is visualised as a marker. Colours of the markers indicate values of the “level of comfort” ratings, with green being comfortable, grey being neutral and red/yellow being uncomfortable

together with people’s emotional responses. This will contribute to a better understanding of “why” people feel comfortable/uncomfortable at particular places.

4 Indoor Mapping (MP3)

4.1 MP3: Organisation and Set-up

Outdoor mapping is today a common task performed by different classes of users, either experts or non-experts. Outdoor open data are currently published: the previously presented OSM is just a relevant example. Standards for data publication and sharing are well defined by Open Geospatial Consortium (http://www.opengeospatial.org/). On the contrary, indoor mapping is an “unexplored field”, with no common or standard tools already defined. Indoor mapping has many applications, both for public utility and for industry: to give a first sight, they span from location-based services in public buildings to unmanned navigation in hospitals, airports and so on. One of these applications is represented by the EU-funded project called i-locate (www.i-locate.eu), which aims at providing a standard framework for applications that allow seamless outdoor–indoor mapping and navigation. The project, ended at the end of 2016, consisted of 18 partners and 13 pilots established all over the Europe.

4.2 MP3: Strategy

During the FOSS4G conference, one of the first attempts (potentially the first) at the organisation of an indoor mapping party was organised to collect data within the Como Campus of Politecnico di Milano. The main purpose was to provide interested people with an opportunity to understand the scientific, technical and practical perspectives of indoor mapping. In short, participants were asked to map indoor spaces of the building by creating rooms inside a GIS application and to literally draw the navigation graph. In order to provide users the necessary knowledge and tools to participate in this party, a brief explanation of the background concepts of the indoor standard (IndoorGML) adopted was provided at the beginning. Together with the basic concepts, two different applications were developed to facilitate the mapping tasks inside the building. The two applications consisted of a desktop and a mobile version that were made available during the party.

The desktop app (Fig. 7) is based on Java OpenStreetMap (JOSM, https://josm.openstreetmap.de). It is an open-source tool, it runs on almost all the desktop operating systems (Windows, Linux and OSX), and it can be installed as a plug-in of the aforementioned tool; in the following, it will be called *i-locate* plug-in. The mobile application provided during the party was an Android application that runs both on phones and tablets, and allows users to easily map the indoor spaces and to create the indoor navigation graph. The i-locate plug-in allows users to draw indoor components (rooms, hallways, elevators and stairs) as basic polygons. These

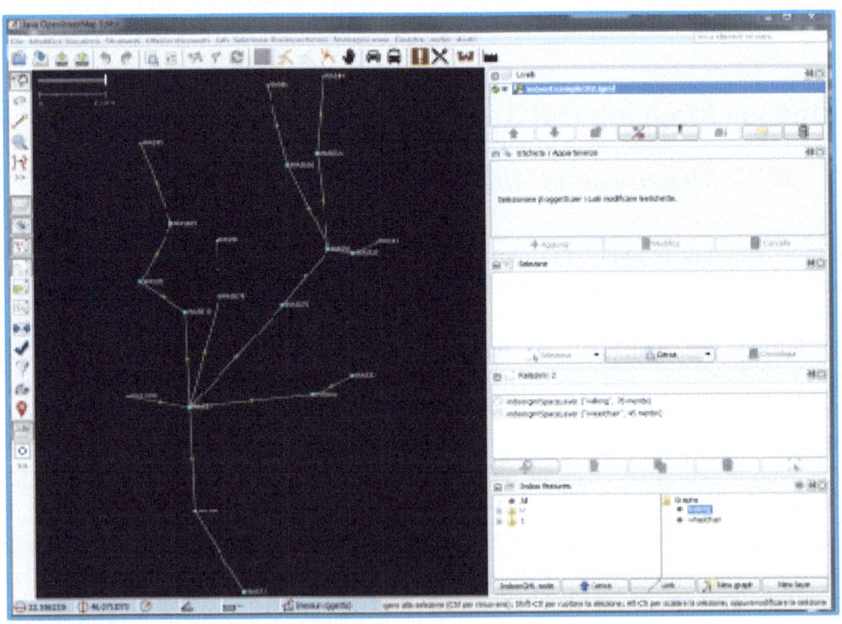

Fig. 7 Desktop app for the indoor mapping party

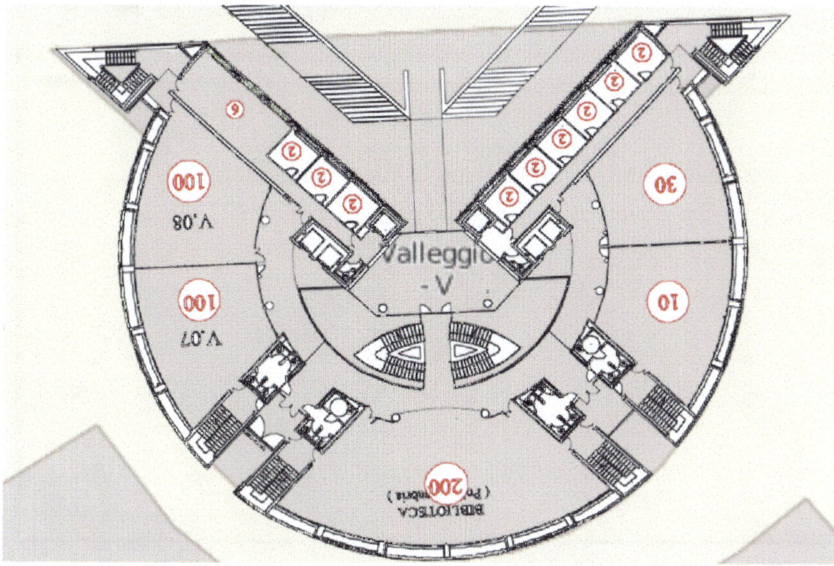

Fig. 8 GeoTIFF map used during the indoor mapping party: it represents the ground floor of one of the buildings (named Valleggio) of Politecnico di Milano at Como Campus

geometries can then be exported as regular geographic data (Shapefiles, GML, GeoJSON and many others) and used in the second step of the mapping. As an alternative, it is possible to import a basemap: at the party, a GeoTIFF (Fig. 8) map has been prepared and provided in order to let users skip the drawings and use it for the "graph mapping".

The IndoorGML standard was used in collecting indoor data in particular to describe the navigation graph (Fig. 9) that users can follow inside the building to reach different destinations. Indeed, in a typical outdoor context, the paths are well defined by streets, sidewalks and so on. On the contrary when indoor rooms are the navigable spaces, that have low or no constraints, it is impossible to have a single and well defined path to follow. However, it is necessary to define and draw all the connections between spaces in order to support navigation systems, and this is what users have achieved during the mapping party.

4.3 *MP3: Participation*

To encourage participation to a new type of mapping party we took a flexible approach to how users would become involved with the mapping party. During the mapping party, some users decided to explore the building before trying to generate data, and some wanted to create the graph while exploring the indoor spaces. They were able to do it by using the mobile application (Fig. 10), which provides similar functionalities to the desktop counterpart, but with easier and simpler tools.

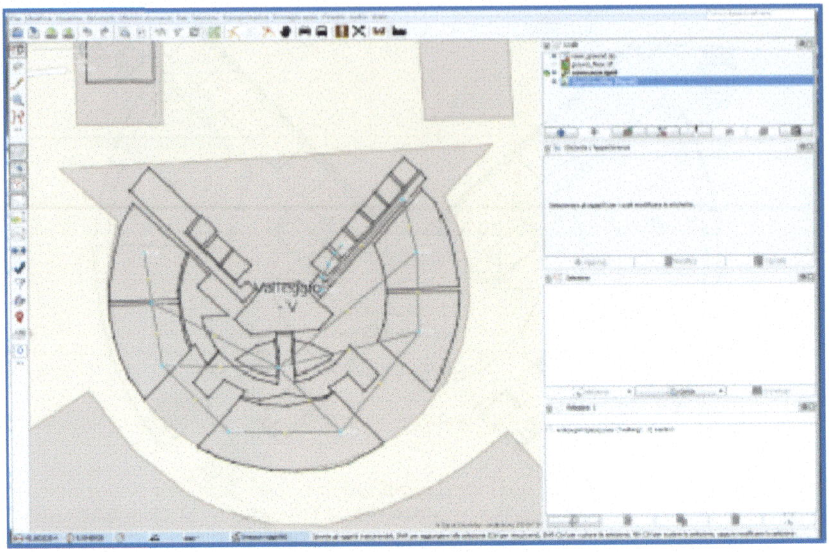

Fig. 9 Desktop app showing the map and the graph of the building

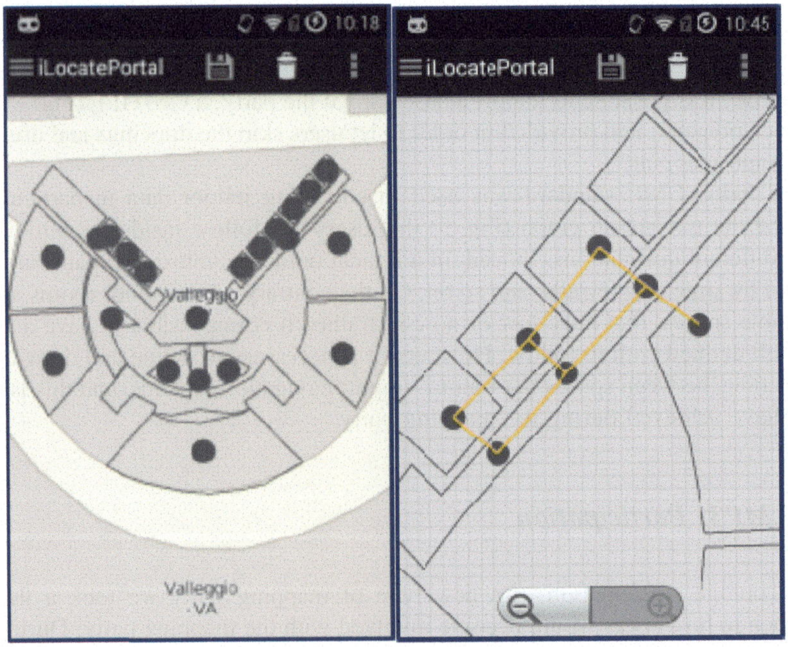

Fig. 10 Mobile application that has been used during the party

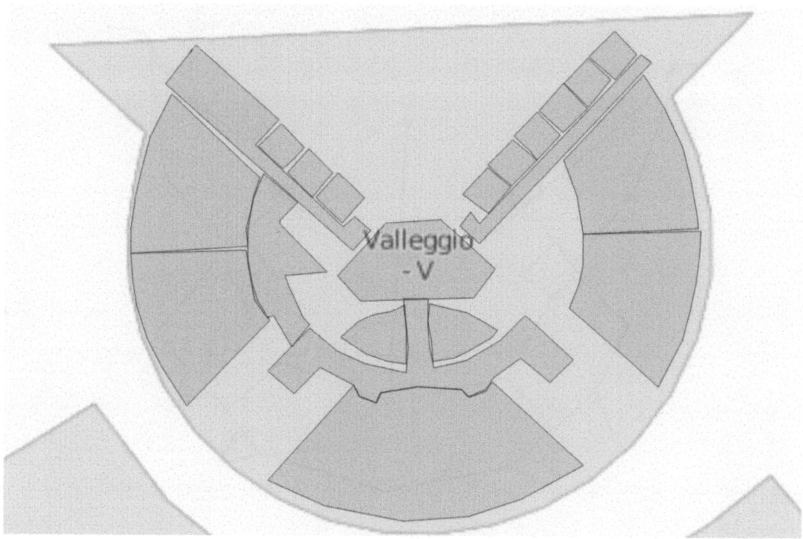

Fig. 11 Merged mapped rooms from the contribution of all the participants in indoor mapping party (example of one roof)

The indoor mapping party received attention from conference attendees and local television and counted more than 30 participants. Almost all the mappers actively generated data but not all of them contributed to the final results. Due to technical limitations, the only way to share the results was to export the data by using the provided applications and send them to the indoor mapping party organisers who merged them (Fig. 11) and generated a single navigable graph (Fig. 12).

4.4 MP3: Results and Outcomes

The mapped building consists of five floors: two of them were very interesting from the point of view of mapping. On average, users have contributed data about twenty rooms per floor per user and an average of 68 indoor navigation nodes per floor per user (with an estimation of 34 edges per user per floor). We felt this was a very good return for the mapping party. As indoor mapping parties are a rather new phenomenon, it is difficult to gauge exactly what the optimal levels of participation or data generation should be.

The collected results constitute a solid base for building an indoor navigation application, which adopts the IndoorGML standard as graph representation (Fig. 13). The experimental phase of indoor mapping environment does not allow to compare the results of this experience with other similar ones, but these results represent an important starting point either as reference or as feedback to improve the next indoor mapping parties.

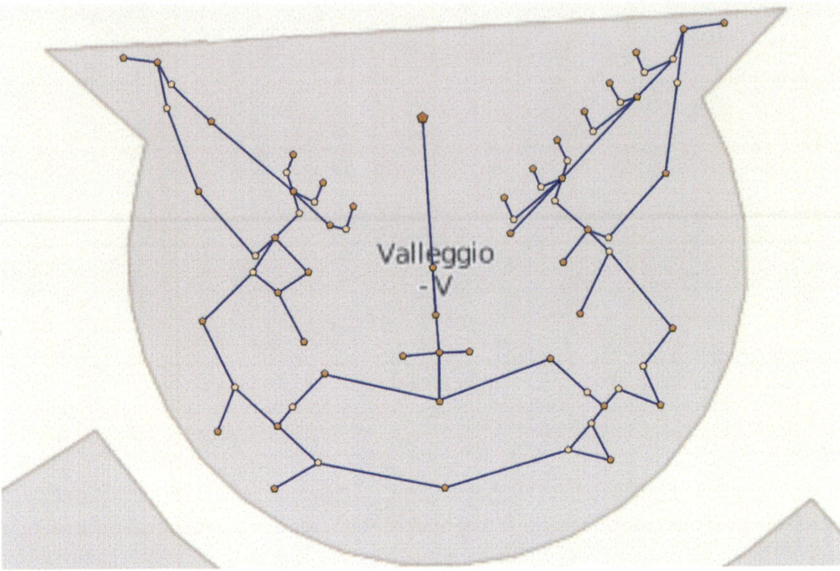

Fig. 12 Merged navigation graph from the contribution of all the participants in indoor mapping party (example of one roof)

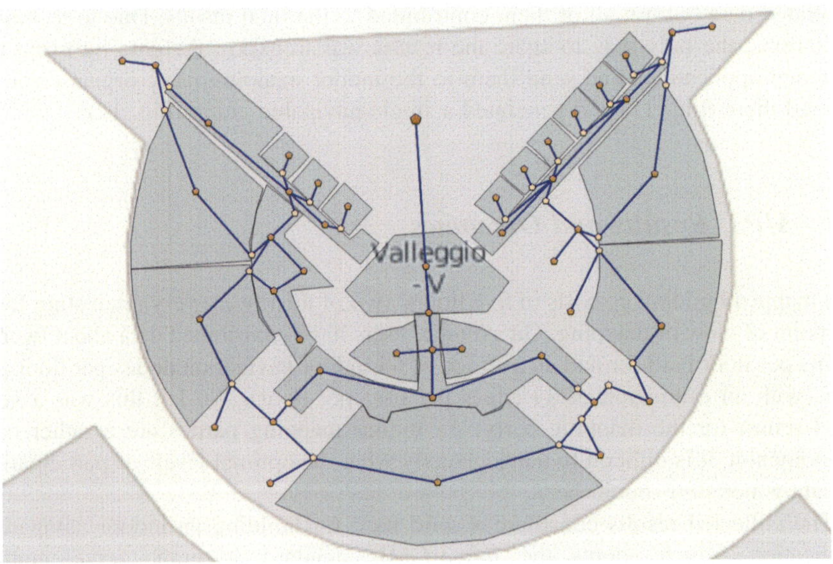

Fig. 13 Final merged result showing both rooms and navigation graph for one roof

4.5 MP3: Lessons Learned and Evaluation

In the organisers' opinion, the indoor mapping party has produced positive results because it acted as an opportunity to obtain feedback for both the mapping community and the indoor mapping group of research. On one side, users have learned new concepts, strategies, problems and tools for allowing people to map and use the indoor spaces in much the same way as we do for the outdoor environment. On the other side, the indoor mapping researchers and developers have received feedback (either positive or negative) regarding the techniques and the tools to generate, publish and use indoor data. This is a strong motivation for engineers to improve the relevant tools and provide more features and usability of their applications. During the months after the party, most of the mappers' issues were addressed: the tools are now easier to use, and features are more powerful. As part of future work, a new mapping party can be organised to test the improvements and to collect more indoor data, which can be shared in the i-locate portal, the first online service to support open indoor data and IndoorGML navigation representation of data.

5 Land Coverage Validation Game (MP4)

5.1 MP4: Organisation and Set-up

The usage of land cover data is very important for different studies related to the environment and sustainable development. Its classification accuracy and validation play a vital role in many applications. The objective of this mapping party is to crowdsource the validation of GlobeLand30, a new global land cover data set at 30 m resolution derived from the classification of Landsat (TM and ETM+) and HJ-1 satellites images according to the pixel-object-knowledge-based (POK-based) approach [13]. A research study evaluated the classification quality of GlobeLand30 on the Italian area [14], in comparison with DUSAF ("Destinazione d'Uso dei Suoli Agricoli e Forestali"), the Italian acronym for "Use Categories of Agricultural and Forest Soil" [15]; the results show a degree of disagreement that ranges between 10 and 20%. Figure 14 shows the distribution of non-coherent pixels for the year 2010 in the Como area. Therefore, another level of validation is needed for the non-coherent data, and we offered this through a Web application in the form of a gaming environment.

5.2 MP4: Strategy

The Land Cover Validation Game [16] is a Game with a Purpose (GWAP) application, as proposed by Von Ahn [17]. A GWAP is a human computation

Fig. 14 Distribution of non-coherent pixels (red areas) in the Como municipality area for year 2010

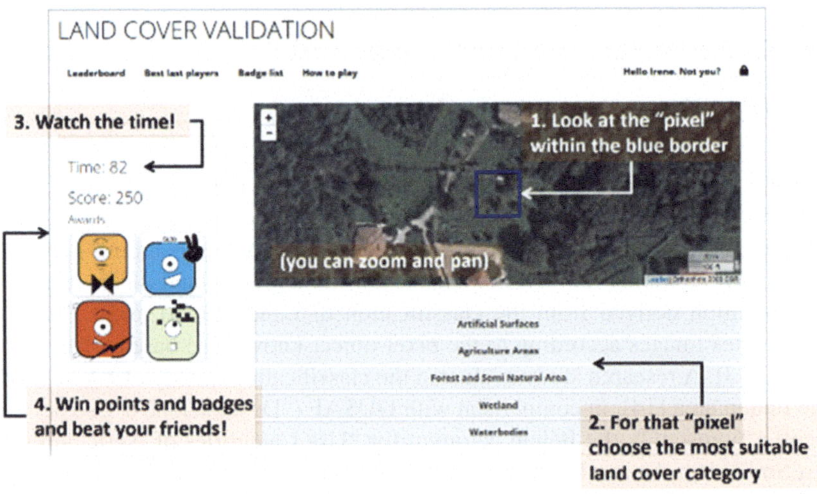

Fig. 15 Land Cover Validation Game interface, with a pixel (blue square box) to be classified

application that puts a task in a crowdsourcing environment with a gaming flavour; the GWAP player has fun playing the game and, as a collateral effect, solves the human computation task. Land cover validation is the purpose incorporated in our GWAP. This interactive game engages the users to classify aerial images for their land use; consequently, the game increases the quality of the land cover classification. Figure 15 shows the user interface of the Land Cover Validation Game: players see an independent orthophotograph (i.e. not used in the classification) of the land under investigation, and they have to classify the "pixel", i.e. a 30 m^2 inside a blue box; depending on the answer, the players get points, badges and level up in a global leader board. While participants classify each pixel, the system collects the answers from multiple players; by cross-checking those replies, the global land cover map is validated, since we can assume that the "correct" land cover for each pixel corresponds to the most "popular" answer. More precisely, the association between each pixel and each of the five land cover categories is annotated with a confidence score between 0 and 1; the land cover types indicated by GlobeLand30 and by DUSAF are given an initial positive score, while the other three are given an initial zero score. Each time a player provides an answer, the confidence information is updated by increasing the score of the association between the pixel and the selected land cover category and by decreasing the other scores. When a score overcomes a specific threshold, the pixel is considered validated and the respective land cover type is associated with the pixel.

5.3 MP4: Participation

The Land Cover Validation Game was advertised at the beginning of the FOSS4G Europe 2015 event during the opening session. The application was made available online at http://bit.ly/foss4game (and it is still available online). Apart from the game intrinsic fun, the participants were given an additional incentive, because an award was foreseen for the top three players in the leader board at the end of the FOSS4G Europe 2015 event. The conference participants played the game during the event week. The Land Cover Validation Game is a Web application; hence, it can be played through any Web-enabled devices; since the game was developed with responsive-design Web technologies, we noticed that, during the FOSS4G Europe event, a significant number of players successfully accessed the game through their mobile phones. During the event week, 68 participants were engaged in the game for a total of more than 20 h of gameplay (average life play of 17.90 min/player).

5.4 MP4: Results and Outcomes

Globally, the 68 participants performed very well. They validated 1600 pixels; therefore, each player on average contributed to the land cover validation of more than 23 pixels. The players' results show an agreement of 86.82% with DUSAF land cover classification, an agreement of 11.87% with GlobeLand30 classification and a disagreement with both classifications only in the 1.31% of cases. A video summarising the Land Cover Validation Game results was presented at the ESA

Earth Observation Open Science event in October 2015, and it is available at https://www.youtube.com/watch?v=Q0ru1hhDM9Q.

5.5 MP4: Lessons Learned and Evaluation

The results outlined above not only confirm the findings of our previous study [16], but also prove that a human computation approach of involving users in a crowdsourcing validation campaign with a gaming incentive is an effective way to collect data and to resolve disagreements between two conflicting land cover classifications. The conference occasion and the natural fun competition between members of the same community definitely constituted a success factor for this user engagement experiment. After the FOSS4G Europe event, during which 1600 pixels from the area of Como were validated, we updated the online Land Cover Validation Game by adding further pixels to be checked by players, both from the same area but related to a different temporal period and from a different area in the city of Milano. Surprisingly, we were able to find that players also voted some classification which is neither DUSAF nor GlobeLand30. This depicts how human computation coupled with the global accuracy assessment of land cover might provide a data set of higher accuracy. Therefore, even if the conference "mapping party" is over, our human computation experiment for land cover classification is still running.

6 Sensing the Conference and the City (MP5)

6.1 MP5: Organisation and Set-up

Together with the more classical mapping party activities, FOSS4G Europe has been complemented with a set of digital services for the audience to enrich their conference participation experience. These services are implemented through a set of technologies that span mobile applications, IoT (Internet of Things) sensors and beacons and social network content analyses and visualisations. While not being a mapping party per se, the aggregation of these technologies and the intensive use that conference participants make of them actually provide a consistent set of information that, when combined together, can provide significant insights into the conference "consumption", spanning all the temporal, geographical and content dimensions. The information collected also allowed the organisers to understand the behaviour of the participants along these same dimensions.

6.2 MP5: Strategy

The first tool provided to the participants was the official mobile app of the conference event. The app is based on the software platform called SocialOmeters,[1] a general-purpose solution created by a joint team of researchers from Politecnico di Milano and developers from WebRatio and Fluxedo (two start-ups founded by Politecnico students and professors). The solution targets organisers of conferences, fairs and events that want to enhance the experience of their participants by providing benefits, information and functionalities through a dedicated mobile app. The assumption is that the event organiser needs a way to provide the participants with detailed information about the event. Based on this, the app features the following capabilities: provision of the official agenda of the event, sessions' details (with time, location, list of presentations, speakers, abstracts and links to additional materials), speaker details, map of the locations, etc. There are also more interactive features such as voting for the different speeches, commenting, social network sharing, message notification and personal agenda setting (where participants can record the list of sessions and speeches they are interested in). The app was customised for the needs of FOSS4G Europe with a dedicated configuration of the event information and programme. Furthermore, the app has been extended with location features, based on positioning of beacons and Bluetooth technology.

6.3 MP5: Participation

Participation in the sensing the conference and city mapping party was encouraged and enabled somewhat differently to the other four mapping parties. In fact, in parallel to the distribution of the app, we worked on the conference venue, by adding instrumentation to it through the dissemination of several *beacons* in the main areas of the building, exploiting the Bluetooth Low Energy (BLE) standard. BLE is a wireless personal area network technology developed by Nokia in 2006 and currently embedded in modern smartphones. Compared to the classical Bluetooth technology, Bluetooth Low Energy is intended to provide considerably reduced power consumption while maintaining a similar communication range. The general term *beacon* refers to small and cheap BLE transmitters, which enable indoor location-based notifications to mobile apps. Beacons transmit small payloads of information, and smartphones listen for the signal transmitted and then respond accordingly. Beacons have been created in order to deliver personalised content directly to a smartphone when it is in the proximity of a specific point. We used this technology for enabling a special feature in the conference app. When a participant entered a conference room during the event, the app immediately notified him and redirected the user interface to the screen showing the details of the conference session/speech that was scheduled for that specific time and place.

[1]http://www.socialometers.com/it/.

6.4 MP5: Results and Outcomes

We equipped the mobile app with the well-known Google Analytics service for monitoring the application usage. This allowed us to extract statistics on the users and the app itself. Some sample statistics are shown in Fig. 16a. The analyses show the total number of users, usage sessions and screen views. It also allows us to define segmentations of the audience based on the language, the phone operating system and brand and so on. In our case, we obtained around 150 users. Main languages were Italian, English, German, Spanish and Dutch. Interestingly, 78% of the users adopted Android systems, with respect to 22% only adopting iOS (iPhone). It is also possible to study the temporal distribution of the usage (e.g. a large peak appeared during the first day of the conference, while usage decreased in the next ones), and the user navigation paths, highlighting the order of visit of the various parts of the application. The latter information also allowed us to determine how many users benefited from the beacons and automatic redirect of the app to the relevant page for the ongoing session. Actually, this happened in just a few cases thus highlighting the fact that Bluetooth is often disabled in devices. It might also indicate that people are still sceptical about automated location-based notifications and information, as they tend to prefer to navigate the app actively. In any case, the analytics for the mobile app provide quite insightful information on the user behaviour, both in terms of when peaks of interest appear and how users explore the information contained in the app. To this respect, Fig. 16b shows the overview of the behaviour flow diagram.

Finally to analyse the behaviour of users, we deployed for FOSS4G Europe and the whole city of Como also a real-time monitor of the social network activities related to the conference [18]. The monitor captured the stream of social network posts on Twitter and Instagram and performed a set of textual and semantic analyses on the content. This allowed the extraction and visualisation of: the geographical position of the posts (when available), the relevant concepts, entities, topics and categories (extracted through semantic analysis), the co-occurrence relations between the topics, as well as the mentions and authorship. All this is exemplified in Fig. 17. Furthermore, the system visualises the temporal distribution of posts, the rankings of top hashtags, entities, users and topics (Fig. 18), discovering emerging concepts, trends and associations [19].

6.5 MP5: Lessons and Evaluation

Globally, during the conference, the system analysed several thousands of contributions and, thanks to the semantic map, was able to highlight trending topics and interactions between users. Overall, the analysis allowed us to understand clearly the geographical distribution of the social network interactions, which concentrated mainly in the conference venue during the main scientific sessions, and then in the

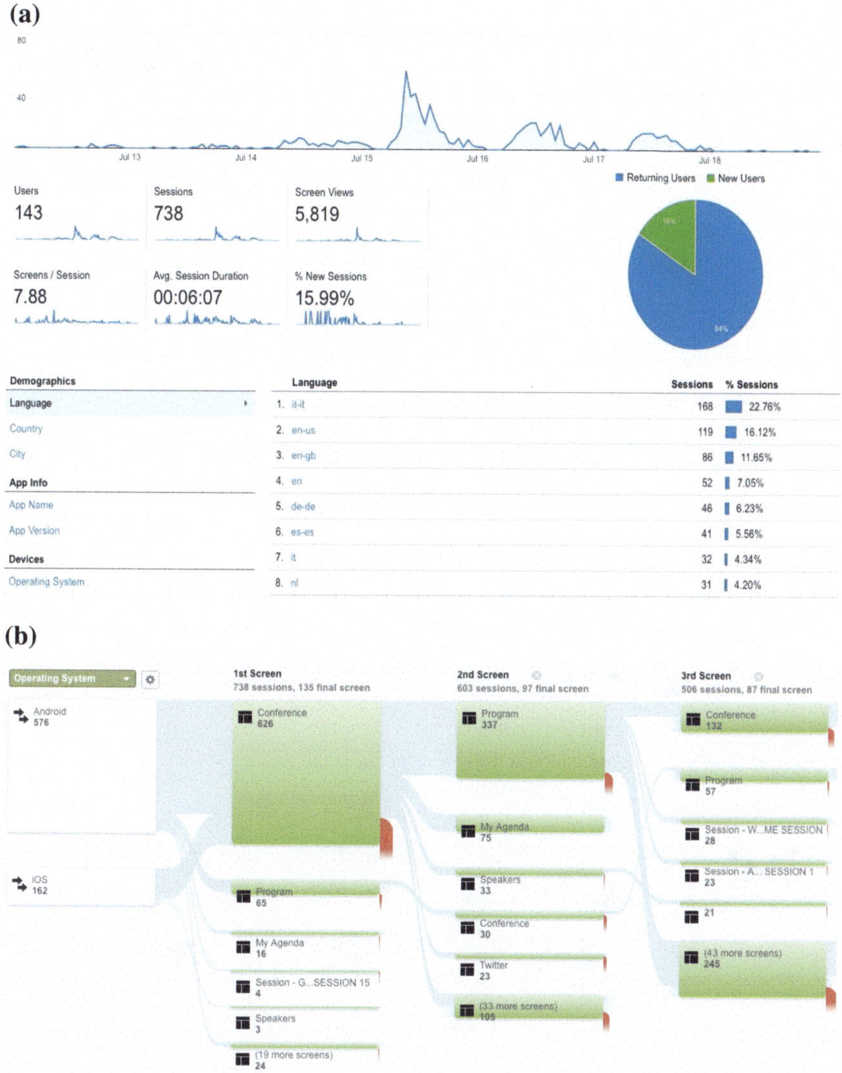

Fig. 16 **a** Set of usage statistics and **b** behaviour flow diagram for the mobile app

touristic and downtown venues in the fringe time slots with respect to the conference programme (morning, late evening and night). As expected, precise geotagging has been much more popular within Instagram posts than with Twitter. This is also due to the specific interaction pattern of the Twitter app, which requires explicit manual confirmation of each geotag.

We foresee a plethora of extensions and challenges to address in the future. First of all, we are currently working on an extension of our data stream analysis and

Fig. 17 Geographical map of geo-located social network posts (a) and semantic map of the discussed topics (b) on Twitter and Instagram

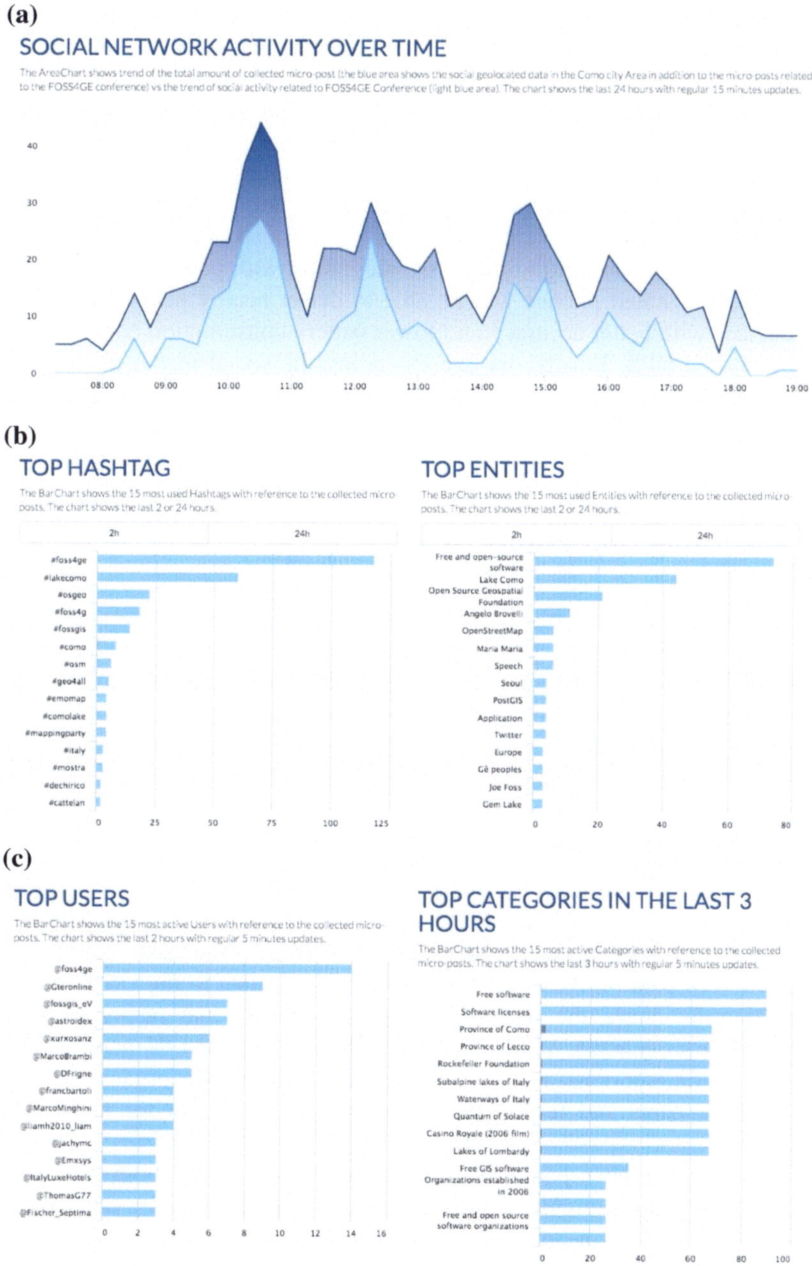

Fig. 18 Temporal distribution of Instagram and Twitter posts during the first conference day (notice the peak during the opening session), together with the rankings of top hashtags, entities, users and topics/categories

visualisation tools that integrate other data sources (beyond social networks), including data from sensors spread at geographical scale across the city. A city-wide experiment is ongoing in Como featuring people counters in the crucial points of the town. Furthermore, we plan to extend the capabilities of the mobile apps by also supporting proximity detection between people. We are currently devising a framework that combines various technologies, such as Bluetooth Low Energy, iBeacon and various sensors of a modern smartphone (compass, accelerometer, pedometer) to create location-based social communities in outdoor and indoor environments, exploiting also integration with personal social network accounts. Instrumented phones will be able to detect nearby people (friends) and smart devices with better accuracy and lower battery consumption compared to the more traditional GPS. Finally, we are extending the social media monitoring by adding additional features, such as measures of "popularity" and "virality" of the analysed content.

7 Conclusions and Future Work

In this paper, we have described the outline, structure, implementation and lessons learned from five mapping parties which were organised during the FOSS4G Europe 2015 conference in Como, Italy. The five mapping parties were as follows: an OpenStreetMap mapping party, EmoMap emotional mapping, indoor mapping, land cover validation and the sensing the conference mapping party. All five of the parties were different in their characteristics: organisation structure, types of technologies used, location (indoor, outdoor, Web-based), etc. Each mapping party learned a number of lessons, and these have been outlined.

- The OpenStreetMap mapping party recommended that such a mapping party should be considered at future FOSS4G events. While the party collected a great deal of extra data for OSM in Como, there were drawbacks related to the collection of data on the street and transfer of this data to the OSM map.
- EmoMap facilitates the reporting of people's emotional responses towards different environments. While the mapping party attracted 94 participants and good data on emotional response to the city of Como, the organisers feel that more participants and responses are required in the future for more detailed analysis.
- The indoor mapping party mapped indoor spaces using both desktop and mobile-based applications. The organisers feel that the mapping party provided a solid foundation for applications such as indoor navigation tools. Since the mapping party, further development on the software tool has been carried out.

Table 1 A summary of the five mapping parties from the preceding discussion of each party

Mapping party	Characteristics	Organisation and structure	Types of techniques used	Locational information	Key lesson or outcome
OSM mapping party (MP1)	Traditional OSM mapping party with the goal to add new information to target area or update existing OSM data	Open call for participation—experienced OSM mappers acted as leaders for groups of mappers. Each group targeted a specific area of Como for mapping. Long process of OSM data upload after the mapping party	Paper-based survey using the Field Papers method. Mappers encouraged to focus on the collection of spatial data and information about POI, roads, buildings, etc. No new geometry collected	All mapping performed outdoors. Survey carried out in Como city centre	When participants learned how to collect OSM data using Field Papers, there was an impressive rate of data collection. This lead to a bottleneck in post-party data upload to OSM
Emotional mapping (MP2)	Use of a mobile location-based application to allow users to specify their emotional response to their current location	Open call for users. Users downloaded the application from the Play Store or Apple App Store. User encouraged to use the app as they moved around Como city	All data captured through the mobile application. All maps displayed at user's current location. Emotional responses selected from predefined lists	The application is designed to capture the emotional responses of users to the outdoor environment. The outdoor location of users is displayed on the app maps	To properly understand people's emotional response to locations, large quantities of data are required. It will be necessary to collect more contextual information about the places on the map
Indoor mapping (MP3)	A desktop GIS application and mobile application were used. Users were required to draw indoor components	Open call for participation. Participants could explore the buildings before beginning data collection	Desktop GIS application and mobile application used. Mapping party organisers assisted with data upload and data sharing. IndoorGML used for data representation	Fully indoor within a multi-floor building.	One of the first mapping parties of its kind. While the overall results and experiences were very positive, there is still some work remaining to improve the tools, test and collect more data and extend to an online service

(continued)

Table 1 (continued)

Mapping party	Characteristics	Organisation and structure	Types of techniques used	Locational information	Key lesson or outcome
Land coverage validation (MP4)	Web-based application using a gamification approach to perform validation of Land Cover Maps	Open call for participation. Participation not limited to conference participants. Application was available for the entire duration of the conference	A simple easy-to-use interface where Web-based maps are used to allow participants to classify land cover for selected pixels	Relates to global land cover mapping. For simplicity, this was restricted to the Como area	Gamification was shown to be a very effective means of engaging participants in the work of land cover validation. However, this experiment continues to operate and collect validations
Sensing the conference (MP5)	Not a traditional mapping party. Considers the collection of social and location-based data from apps, social media and physical infrastructure such as beacons	Available at all times during the conference week to "sense the conference and the city". Official conference navigation app encouraged delegates to use the app	Passive, non-invasive sensing and collection of data from delegates' use of the conference app, interactions on social media, delivery of personalised location-based content	Real-time monitoring of social media activities during the conference week. Office conference mobile application. Not necessarily strictly indoor or outdoor	Indications of the ability in the future to create location-based social communities in both indoor and outdoor environments and activities

- The Land Coverage Validation Game/Party sought to crowdsource the validation of the GlobeLand30 data set. A prize was awarded for the top three players or validators in the game. This was the only party which offered this incentive for participation. The organisers believe that the combination of crowdsourcing and a gaming-based incentive is an effective way of collecting data and classifications for land cover.
- The sensing the conference was not an actual mapping party. It was a software-based infrastructure which analysed social network activity about the conference. It also included a mobile application which participants used as they attended the conference. The organisers believe that this approach to sensing a conference has a very exciting future with extensions to social data stream analysis and visualisation planned.

To summarise the mapping parties, we provide a summary table in Table 1. In the table, there are five columns. *Characteristics* gives a brief overview of the aims of the corresponding mapping party. *Organisation and Structure* summarises how the mapping party was set-up. *Types of Techniques used* summarises the technological approaches used in the mapping party to capture data or generate information. *Locational information* indicates if the mapping party took part indoors, outdoors or the specific spatial region targeted by the mapping party. The final column *Key Lesson or Outcome* provides one key lesson or outcome from the mapping party summarised from the details given above.

Together with Table 1, Table 2 allows for a quick visual comparison of all mapping parties under several different headings. Overall, the mapping parties provided an environment of fun, mapping and engagement among the conference participants. All of the mapping parties were organised as free time events for participants or as part of Web-based or mobile-based applications. While there were campaigns for recruitment of participants overall any participation was voluntary. The key messages from these mapping parties were that the conference provided a very suitable environment for geographical and software-application themed mapping parties. The recipe to success was found through (1) active recruitment of participants, (2) low or no technical or skills barriers to participation, (3) the appropriate use of GIS software and technology and (4) a fun and relaxed environment. All of the mapping party organisers agreed that it is difficult to precisely indicate how to further increase participation in each mapping party for a conference-based deployment. Indeed, we felt that overall participation was strong. The conference environment is a busy and time-sensitive one. Conference delegates arrive with different goals, perceptions, levels of willingness to participate in conference activities, willingness to engage socially with other delegates, etc. The obvious next step we shall consider is the organisation of these mapping parties, or similar, at future FOSS4G events with the benefit of the knowledge and lessons learned from FOSS4G Europe 2015. There is certainly scope to consider a cross-examination of the outputs and data generated by each of the mapping parties. This type of an analysis could have multiple aspects. For example, it could investigate if specific objects, areas, regions, etc., were of particular interest across

Table 2 A summary of the participation and contribution levels for each mapping party

Mapping party	Participation and contribution summary
OSM mapping party (MP1)	Forty participants (16 M, 24F), approximately 700 new nodes added to Como OSM
Emotional mapping (MP2)	Ninety-four participants, 244 emotional responses, mean contributions per participant was 2.59 responses
Indoor mapping (MP3)	More than 30 people. On average 68 indoor nodes per floor per user, 34 edges per user per floor
Land coverage validation (MP4)	Sixty-eight participants were engaged in the game for a total of more than 20 h of gameplay (average life play of 17.90 min/player). Validation of 1600 pixels. Average validation 23 pixels
Sensing the conference (MP5)	Around 150 users. No actual contributions. Rather passive monitoring of social media contributions, app usage, etc.

all of the mapping parties. It could also consider the levels of user contribution across all parties—are some users participating in all parties or just one.

However, in order to properly extend the findings and outcomes of these mapping parties it is acknowledged that a much wider participation is required and within different participation environments. Within the FOSS4G Europe conference setting, all of the mapping parties connected directly with an audience who were technically skilled, comfortable with using spatial data and spatial data technologies, had free time available to participate and had mobile or computing devices available to make them capable of using the mapping party software, etc. There was also a synergistic effect in having five mapping parties running simultaneously during the conference. This provided incentives for each of the individual mapping party leaders to actively advertise their mapping party and recruit volunteers. Anecdotally, it also provided an incentive for the conference delegates to become involved in at least one mapping party for a variety of reasons including having fun, helping colleagues with their research, pursuing applications of interest to them. In many ways, these scenarios paint an artificial and unequal picture of participation which we would expect to be very different outside the confines of the conference environment.

Conferences such as FOSS4G and thousands of other large ICT conferences held around the globe annually often situate themselves within a city or region without actually physically interacting with the host city or its citizens. The conference does not necessarily give back anything to the city and its residents nor does it leave the city with any tangible benefits, with the obvious exception of tourism revenue, of having hosted the conference. Overall, we feel that the mapping parties described here generated very positive and tangible benefits for the city of Como. The emotional mapping party saw the presence of citizens of Como as well as conference delegates, and this party was well advertised in the days before the conference in local media. The land coverage game specifically used the host city as the geospatial objects for validation. This gave, in principle, the possibility for players to verify classifications and in turn gave back to the city a better overall classification of land cover validated during the game. The OSM had a similar effect in improving the OSM map of Como. The mapping parties can therefore give delegates the possibility of getting to know the host city better while providing a platform upon which these delegates can give back something to the host city—in this case better geospatial information about it—which will benefit every resident of the host city. We argue that this transforms the conference from an isolated event closed to the public and the residents of the host city to an event which interacts with the city. This generates tangible and intangible benefits and values in hosting these kinds of conference events. These types of contributions to the host city are to be celebrated as new and significant outcomes and should be considered as a part of future FOSS4G and ICT conferences.

Acknowledgements We want to warmly thank all participants to FOSS4G Europe 2015 who were so enthusiastically involved in all the mapping parties, giving us also the possibility of testing these new collaborative ways of collecting geospatial data. The support of EU COST Action IC1203 "European Network Exploring Research into Geospatial Information Crowdsourcing: software and methodologies for harnessing geographic information from the crowd (ENERGIC)" is also gratefully acknowledged.

References

1. Goodchild, M. F. (2007). Citizens as sensors: The world of volunteered geography. *GeoJournal, 69*(4), 211–221.
2. Neis, P., & Zipf, A. (2012). Analyzing the contributor activity of a volunteered geographic information project: The case of OpenStreetMap. *ISPRS International Journal of GeoInformation, 1,* 146–165.
3. Soden, R., & Palen, L. (2014). From crowdsourced mapping to community mapping: The post-earthquake work of OpenStreetMap Haiti. In: *Proceedings 11th International Conference on the Design of Cooperative Systems (COOP 2014), May 27–30 2014, Nice, France* (pp. 311–326). Springer International Publishing.
4. Haklay, M., & Weber, P. (2008). OpenStreetMap: User-generated street maps. *IEEE Pervasive Computing, 7*(4), 12–18.
5. Perkins, C., & Dodge, M. (2008). The potential of user-generated cartography: A case study of the OpenStreetMap project and Manchester mapping party. *North West Geography, 8*(1), 19–32.
6. Hristova, D., Quattrone, G., Mashhadi, A. J. & Capra, L. (2013). The life of the party: Impact of social mapping in OpenStreetMap. In: *Proceedings 7th AAAI Conference on Weblogs and Social Media (ICWSM 2013), July 8–11 2013, Boston, MA, USA*.
7. Mooney, P., Minghini, M., & Stanley-Jones, F. (2015). Observations on an OpenStreetMap mapping party organised as a social event during an open source GIS conference. *International Journal of Spatial Data Infrastructures Research, 10,* 138–150.
8. Huang, H., Klettner, S., Schmidt, M., Gartner, G., Leitinger, S., Wagner, A., et al. (2014). AffectRoute – Considering people's affective responses to environments for enhancing route planning services. *International Journal of Geographical Information Science, 28*(12), 2456–2473.
9. Barrett, L. (2006). Valence is a basic building block of emotional life. *Journal of Research in Personality, 40*(1), 35–55.
10. Russell, J. (2003). Core affect and the psychological construction of emotion. *Psychological Review, 110*(1), 145–172.

11. Spencer, E. H. (2015). How to analyze Likert and other rating scale data. *Currents in Pharmacy Teaching and Learning*, *7*(6), 836–850. ISSN 1877–1297, http://dx.doi.org/10.1016/j.cptl.2015.08.001.
12. Russell, J., & Pratt, G. (1980). A description of the affective quality attributed to environments. *Journal of Personality and Social Psychology, 38*(2), 311–322.
13. Chen, J., Chen, J., Liao, A., Cao, X., Chen, L., Chen, X., et al. (2014). *Global land cover mapping at 30 m resolution: A POK-based operational approach.* Remote Sensing: ISPRS J. Photogram.
14. Brovelli, M. A., Molinari, M. E., Hussein, E., Chen, J., & Li, R. (2015). The first comprehensive accuracy assessment of GlobeLand30 at a national level: Methodology and results. *Remote Sensing, 7,* 4191–4212.
15. Credali, M., Fasolini, D., Minnella, L., Pedrazzini, L., Peggion, M., & Pezzoli, S. (2011). Tools for territorial knowledge and government. In: D. Fasolini, S. Pezzoli, V. M. Sale, M. Cesca, S. Coff ani, & S. Brenna (Eds.), *Land cover changes in Lombardy over the last 50 years* (pp. 17–19). Milano: ERSAF.
16. Brovelli M. A., Celino I., Molinari M. E., & Venkatachalam V. (2015). Land cover validation game. *Geomatics Workbooks no. 12 – FOSS4G Europe Como 2015* (pp. 153–157).
17. Von Ahn, L. (2006). Game with a purpose. *IEEE Computer, 39*(6), 92–94.
18. Arnaboldi, M., Brambilla, M., Cassottana, B., Ciuccarelli, P., Ripamonti, D., Vantini, S., & Volonterio, R. (2016). Studying multicultural diversity of cities and neighborhoods through social media language detection. In: *CityLab Workshops of the Tenth International AAAI Conference on Web and Social Media (ICWSM 2016), Cologne, Germany*. AAAI Press, Palo Alto, California. ISBN 978-1-57735-768-1.
19. Brambilla, M., Ceri, S., Della Valle, E., Volonterio, R., & Acero Salazar, F. X. (in press). Extracting emerging knowledge from social media. In: *26th International World Wide Web Conference (WWW 2017), April 2017, Perth, Australia.*

Crowdsourcing to Enhance Insights from Satellite Observations

Suvodeep Mazumdar, Stuart N. Wrigley, Fabio Ciravegna, Camille Pelloquin, Sam Chapman, Laura De Vendictis, Domenico Grandoni, Michele Ferri and Luca Bolognini

Abstract Insights from satellite observations are increasingly being used to enhance a range of domains from highly specialised scientific research through to everyday applications directly benefiting members of the public. A particular category of satellite observations—Earth Observations (EO)—is concerned with capturing information regarding the Earth's atmospheric and environmental conditions and observing human activity and its impact on the Earth's surface. A growing number of technologies and services heavily rely on EO data and the rapidly improving fidelity, coverage, timeliness and accessibility of such observations are providing significant opportunities for new applications of economic and societal benefit. With the increasing importance, relevance and size of EO data sets, it is critical to understand how the value of such data can be maximised by complementing EO with other sources of data and efficiently making complex interpretations and decisions. The wide adoption and availability of smartphones, Internet devices and increased

S. Mazumdar (✉)
Department of Computing, Sheffield Hallam University, Sheffield, UK
e-mail: s.mazumdar@shu.ac.uk

S. N. Wrigley · F. Ciravegna
Department of Computer Science, University of Sheffield, Sheffield, UK

S. N. Wrigley
Department of Automatic Control and Systems Engineering,
University of Sheffield, Sheffield, UK

C. Pelloquin
Starlab Ltd., Oxford, UK

S. Chapman
The Floow Ltd., Sheffield, UK

L. De Vendictis · D. Grandoni
e-GEOS S.p.A, Rome, Italy

M. Ferri
Alto Adriatico Water Authority, Venice, Italy

L. Bolognini
aizoOn Technology Consulting, Milan, Italy

G. Bordogna and P. Carrara (eds.), *Mobile Information Systems Leveraging Volunteered Geographic Information for Earth Observation*, Earth Systems Data and Models 4, https://doi.org/10.1007/978-3-319-70878-2_2

accessibility to information has paved the way for large numbers of citizens and communities to participate in scientific, technological, societal and decision-making activities. This chapter discusses the experience of the European Space Agency funded Crowd4Sat project led by the University of Sheffield that investigated different facets of how crowdsourcing and citizen science impact upon the validation, use and enhancement of Observations from Satellites products and services.

1 Introduction

The role of citizen science (CS) and crowdsourcing is vital to a wide range of applications, spanning a large number of fields such as science, governance, public policy, environmental studies and decision-making. Citizens have been employed in scientific studies and decision-making processes over the years and several excellent examples have showcased how citizen-generated data can provide high-quality data. Although concerns exist regarding the assessment of quality and reliability of CS data [1, 2], several domains such as knowledge bases, mapping, classification have demonstrated high quality achieved through the rigour of CS combined with multiple independent reviews to check reliability. It has also been reported that such data can be more detailed and higher quality than provided by official institutions [3–6]. Several large organisations such as Amazon, Trip Advisor, Twitter and Facebook also rely on crowdsourcing as primary sources of information, comprising a critical aspect of their entire business model. Wikipedia and OpenStreetMap, on the other hand serve as long-standing testament to the provision of open data which is created, maintained and enriched by the public. With the potential of engaging with citizens, it is important to study how the value of their contribution can be maximised most effectively. The domain of Earth Observation (EO) is increasingly employing CS and crowdsourcing for tasks such as calibration and validation of data as demonstrated by the growing number of publications in the field [7]. While the potential of CS applied in EO can be immense, it is important to understand various factors in engaging citizens and exploiting their contributions in an operational context. There is also a need to study different approaches of employing citizens in different settings and tasks in order to develop techniques and mechanisms for effectively using crowdsourced data. Finally, it is important to understand the aspects needed to be considered while developing crowdsourcing solutions for EO, from the perspective of different stakeholders such as authorities, decision-makers, researchers, industry.

The Crowd4Sat project[1] led by the University of Sheffield was a study funded by European Space Agency (ESA) and informed by demonstration projects which investigated how CS and crowdsourcing could contribute to the enhancement, use and validation of satellite observation and products. The project explored a range of

[1]http://www.crowd4sat.eu.

crowdsourcing methodologies and technologies, from opportunistic sourcing (the ability to extract relevant data from unrelated activities) to participatory sourcing (citizens and authorities explicitly participate in data collection). The crowdsourced data and information collected ranged from real-time vehicle mobility telemetry (via the The Floow's technologies used by hundreds of thousands of users on four continents) which was analysed to estimate geospatial vehicular pollution, to crowdsourced geotagged images (from sites such as Panoramio, Flicker and Twitter) and social media messages (e.g. from Facebook, Twitter, Foursquare). Observations from Satellites (OS) data were sourced from a wide range of ESA missions and products including ERS[2]/Envisat[3] and Sentinel-1[4] and additional OS sources such as Landsat-8[5] and MODIS.[6] A variety of stakeholders such as authorities, emergency responders, city councils, insurance companies, as well as individuals and citizen associations have been involved throughout the process. The project addressed concrete scientific and societal problems through four use cases demonstration projects, targeting key scientific and societal issues: pollution in metropolitan areas; land use; water management and snow coverage; and flood management and prevention. The Crowd4Sat project kicked off on February 2015 and over the duration of 14 months comprised of two main strands: strategic roadmapping, understanding the state of the art and demonstration projects. A variety of activities were conducted throughout the project including conducting reviews of relevant initiatives, roadmapping activity, stakeholder analysis, requirements analysis, technology design and development, user evaluations and stakeholder feedback. The work assessed the feasibility of adding value to ESA space products and services by using crowdsourcing and citizen science by understanding the practical limitations and issues that can arise out of engaging with citizens and communities. Several recommendations were identified for ESA and citizen science communities.

This chapter introduces the demonstration projects and discusses the findings from the project, to provide recommendations for developing tools and technologies for crowdsourcing for EO.

2 Citizen Engagement and Participation in Citizen Science

Several definitions of 'citizen science' have been proposed, the earliest being by Irwin [8], describing how citizens accumulate knowledge in order to learn and respond to environmental threats. References [9, 10] refer to 'citizen science' as a form of research collaboration to address real-world problems. Reference [1]

[2]https://earth.esa.int/web/guest/missions/esa-operational-eo-missions/ers.

[3]https://earth.esa.int/web/guest/missions/esa-operational-eo-missions/envisat.

[4]https://earth.esa.int/web/guest/missions/esa-operational-eo-missions/sentinel-1.

[5]https://landsat.gsfc.nasa.gov/landsat-8/.

[6]https://modis.gsfc.nasa.gov/about/.

defined citizen science as scientific activities that non-professional scientists volunteer to participate in data collection, analysis and dissemination of scientific projects. While being only recently coined as a term in a variety of nuances as "a collaboration" [10], "research tool" [11], "genre of (mobile) computing" [12], "activities" [1] and a "trend" [13], the engagement of public in professional research and activities has had a long history dating back to over two centuries. In essence, the field of citizen science has transitioned from citizens having alternate sources of employment conducting scientific activities out of interest to citizens being employed in institutions for conducting research as professional scientists. This transition to professional scientists observed a major growth since the latter part of the nineteenth century [1], with increased institutionalisation of science and scientific activities. With the establishment of organisations, research institutions and universities, research activity itself has undergone a tremendous transformation, underpinned by scientific rigour, processes and protocols. While this has, in many ways contributed to the alienation of citizen scientists, the role of citizens as contributors has continued throughout this process, albeit in selected areas of study such as archaeology, ecology and natural sciences. Citizen roles, for such areas, have mainly focussed on the process of data collection and cataloguing observations. Citizens now have ever increasing means of contributing to citizen science, with the smartphone industry revolutionising how citizens can provide data: actively (explicitly sending information via mobile applications, Websites, etc.) and passively (collecting and sending data without an active involvement of citizens). Also, in addition to providing observations and opinions in a standardised and processable manner, citizens can also provide evidence for their observations by submitting media (images, audio, video), along with metadata (e.g. timestamps, geolocation, exchangeable image file format). At the same time, hobbyists and enthusiasts can build their own low-cost physical environmental sensors, which can be easily connected to sensor networks via Application Programming Interfaces (APIs). Similar sensors can also be bought off the shelf and distributed to communities to be deployed in larger areas than covered by highly expensive, professional sensors traditionally provided by local, regional or national authorities or agencies. All of these approaches eventually contribute to a greater awareness of the environmental and physical conditions, with a far wider coverage than previously possible. With the availability of social media and qualitative views of citizens, situations on the ground can be far better understood than before.

A key technical component of citizen science is the process of collecting information from citizens via crowdsourcing. The term crowdsourcing was originally coined by Jeff Howe, contributing editor for Wired Magazine, as a portmanteau of outsourcing and crowd [14]. The Oxford English Dictionary defines crowdsourcing as '*the practice of obtaining information or services by soliciting input from a large number of people, typically* via *Internet and often without offering compensation*'. Although this definition mentions use of the Internet, this does not have to be the case—tasks can be offline as well as online. Indeed, some of the earliest examples of crowdsourcing for citizen science have been offline processes [15] where participants sent information and photographs to researchers via postcards. With the availability of different forms of technologies and increasing possibilities of contributing with information, citizens can now participate in a variety of mechanisms based on their levels of engagement—they can merely provide access to computational resources or be highly active, performing tasks and collecting data. Broadly, citizen science projects have been classified by [16] according to increasing engagement by participant as follows:

2.1 Passive Sensing

In passive sensing, participants allow data generated by equipment owned by themselves (e.g. mobile phones, environmental sensors, GPS units) to be collected and used by researchers. Upon initial setup and configuration, most often the volunteers are not expected to actively participate in the project, and their data is seamlessly collected in the background. For example, the Weather Underground project has a network of over 100,000 personal weather stations across the US, combining with federally funded 'official' weather stations to provide data for forecasting.

2.2 Volunteer Computing

In volunteer computing, participants provide spare computing resources to enable researchers to create a 'virtual compute grid', enabling larger amounts of data to be processed more economically than would be possible using local compute resources. One of the first examples of volunteer computing is the large SETI@Home project, launched in 1999 as an Internet-based project. In the project, volunteers install a program that downloads and analyses radio telescope data with the purpose of detecting intelligent life outside Earth.

2.3 Volunteer Thinking

Also referred to as Human Intelligence Tasks (HIT), volunteer thinking involves participants donating some of their spare time to perform some tasks such as data analysis, visual observations, annotation. These tasks are usually performed online using computers, tablets or smartphones, mostly requiring classification of images or recognition of patterns. Typically, larger jobs are split into smaller tasks that are distributed to large number of workers who, with little effort, contribute towards solving a larger task. Microtasks are generally tasks that are difficult for computers and algorithms to complete accurately such as audio transcription and handwriting recognition. The solutions are often used as training data and ground truth for machine learning algorithms to help improve the performance of automated audio transcribers, etc. At the beginning, volunteers are usually asked to undertake some basic training to understand how to correctly perform the tasks. Overall, the amount of time required is low and individual tasks can be quickly completed. One of the most widely adopted and popular examples of volunteered thinking is Galaxy Zoo [17], where volunteers are involved in morphological classification of galaxies from images. Websites such as Amazon Mechanical Turk and CrowdFlower provide platforms for citizen science projects to employ microworkers and define tasks and rewards which can then be made available for crowdworkers to search and select topics of interest to them. Several citizen science projects have also employed crowdworkers to solve tasks such as transcribing historical documents [18], detecting colorectal cancer polyps in image scans [19], text annotation of medical documents [20].

2.4 Environmental and Ecological Observation

Environmental and ecological observations involve participants in monitoring and observing the environment for different purposes. Many of the most established examples of citizen science and crowdsourcing fall into this category. For example, the National Audubon Society has been conducting the Christmas Bird Count annually since 1900, involving collecting observations from thousands of participants globally. In the UK, the British Trust for Ornithology conducts similar surveys.

2.5 Participatory Sensing

Engaged with a higher degree of control and influence over the data collected and analyses, participatory sensing enables volunteers with a greater amount of participation. Activities are typically initiated by external research organisations with close cooperation with volunteers. Such activities typically exploit advanced technologies available in mobile phones. Some examples of participatory sensing include air quality sensing [21], noise level sensing via mobile's microphone readings and geographical locations [22].

2.6 Civic/Community Science

Initiating from the citizens and communities themselves, civic science is initiatives where many (if not all) stages of the scientific processes are conducted by non-professional scientists. For example, residents on the Pepys Estate in London concerned about air quality and pollutants released into the atmosphere by a local scrapyard initiated and guided data collection from citizens. External research organisations were also contracted to perform more specific analyses.

Passive sensing and volunteer computing, while being simpler ways of engaging citizens in science and collect data which are relatively trustworthy, their engagement, and often cognitive capability, is wasted. At the same time, a larger participation from citizens can be possible particularly for volunteers who either do not have enough time to commit to physical or cognitive effort, or those with lower educational attainment. Volunteer thinking and environmental and ecological observation are usually the most commonly deployed and historically established forms of crowdsourcing. Compared to environmental and ecological observation, volunteer thinking, however, expects a much lower commitment from participants in terms of physical engagement: volunteers are not expected to take readings by visiting locations and carry equipment, etc. Instead, volunteer thinking tends to involve rapid or widespread data collection/analysis—e.g. Galaxy Zoo [17] involves the rapid assessment and categorisation of images of galaxies. Participatory sensing and civic science heavily rely on significant engagement from the volunteers, not just in the collection and interpretation of collected data but also in organisation, management and curation of the project. Particularly for civic science, citizens and professional scientists collaborate as peers in all stages of the scientific process.

3 Stakeholders in the EO Domain

The success of citizen science and crowdsourcing often relies on understanding the various actors who are involved in the process—not only in data collection, but also analysis, project management and end users. This is a highly important aspect as it helps understanding various expectations from different types of stakeholders. Stakeholders are, in this case, the groups potentially interested in crowdsourcing activities and the results originating from them. In the context of the Earth Observation domain, the Crowd4Sat project identified several groups of different nature: decision-makers, scientific teams, industry and citizens. As data from OS does not always meet the expectation of stakeholders in terms of spatial and temporal resolution or information need, such groups have a significant interest in understanding how crowdsourcing can help in improving, validating and extending such data sets. Decision-makers are in general highly interested in citizen engagement and crowdsourcing as a source of data as well as gathering awareness on key societal issues. In addition to understanding local issues and concerns, authorities also rely on highly accurate information such as OS maps to support planning activities and investigations. Through a large-scale citizen participation, crowdsourced data can provide them with in situ measurements that can validate OS and fill-in various gaps identified in such data sets. At the same time, engaging with citizens can help authorities understand highly evolving situations on the ground. For example, during emergencies, citizens tend to be the first observers of unexpected events and hence can provide highly accurate real-time information on events. Crowdsourcing and citizen science enable citizens to become highly involved and engaged actors in decision-making processes, eventually helping authorities make decisions better understood by citizens and better aligned with community's interests and concerns. In spite of the improved situational awareness, encouraging citizen participation by authorities is a potentially high-risk activity. It is important to consider several factors in such scenarios, for example, ensuring citizens are not put in harm's way as a result of their interest in helping authorities during emergencies. Furthermore, authorities need to also consider the responsibilities and implications of citizen participation such as the need to act upon all information arising out of citizens during large-scale events.

Various economic factors also drive the need for improving OS via crowdsourcing for the industry. For example, various companies often look for extended coverage and accurate measurements for resource estimation. With massive crowdsourcing activities, such data can provide in situ measurements to validate OS products at costs far lower than traditional surveys or data collection activities. For instance, insurance providers need to evaluate risks of natural disasters and limit potential losses that may arise as a result. Such providers can be assisted by improved OS quality by using crowdsourcing using more accurate and up-to-date measurements. The scientific community now faces significant challenges due to funding constraints and can hence benefit from large pools of volunteers who could contribute by either providing in situ measurements, validating observations or even conducting analyses or performing tasks. Many researchers rely on crowdsourcing to provide essential data for their research. Citizens can also benefit from crowdsourcing either personally or via citizen associations (e.g. hikers associations, bird-watching associations, local action groups). Citizen data, in large amounts, can bring interesting benefits to citizens themselves through better understanding of their environment, activism around local issues, assisting in existing activities and hobbies and so on. Furthermore, several solutions exist that exploit citizen-generated data and are made available to consumers as products. For example, Google Traffic uses passively crowdsourced mobile phone traces to estimate traffic conditions which are subsequently used for journey planning.

4 Demonstration Projects

Crowd4Sat addressed key scientific and societal problems through four demonstration projects (DPs) by combining OS with crowdsourced observations. Each DP had a set of stakeholders who were approached to gather initial sets of requirements, which helped set a clear focus on the user needs to ensure a profitable uptake of the products and services offered by the project. The process of requirements gathering was conducted via user meetings during the first three months of the project. DP1 was aimed at validating snow coverage maps produced from MODIS and Sentinel products with crowdsourced information collected through a dedicated mobile application, distributed within the hiking community in Catalonia, Spain. This activity involved interviewing stakeholders from multiple organisations: Federació d'Entitats Excursionistes de Catalunya (Hikers Association), Agéncia Catalana de l'aigua (Water Agency) and Asociación turística de estaciones de esquí y montaña (Ski Resorts Association). The monitoring of snow melt is a key parameter for management of water resources and runoff modelling. In this context, OS is very useful and has reached operational maturity. Furthermore, new satellites such as ESA's Sentinel-1A are expected to improve monitoring of snow cover areas with greater accuracy and improved revisit times. However, measuring snow cover areas with SAR satellite image processing comes with intrinsic slant-range distortion problems such as foreshortening, layover, shadowing. This DP proposed to exploit crowdsourced observations from hikers to validate processed snow coverage maps. Hikers were approached via hiking associations and were provided with Android and iOS mobile applications, which would allow sending reports of snow presence/absence information (along with geo-localised images) to a server collecting all information. The mobile apps would also provide hikers with hiking route tracks to provide further means of engagement.

DP2 was aimed at understanding how opportunistically crowdsourced vehicle telematics data can be combined with satellite remote sensing and in situ data to improve pollution mapping and modelling for local authorities of large metropolitan areas. The interviewees for this activity were primarily Sheffield City Council (Traffic and Planning, Air Quality, Strategy, Sustainability), South Yorkshire Transport Planning Executive body and South Yorkshire Intelligent Transport Systems. This DP aimed at enabling usage of mass road crowdsourced mobility data to help better understand road pollution. Existing methodologies to observe road pollution involve few ground-based calibrated monitoring stations, but such sensors fall short of providing fine-grained emissions across road networks. Such sensors are only few, owing to the high cost of procurement and installation as well as located in locations not ideal due to the need for installation in safe places. The project seeked to address this data gap by using a combination of crowdsourced mass mobility GPS trace data, Corine OS Land Usage data and digital elevation models.

DP3 evaluated how opportunistically crowdsourced Social data (e.g. Twitter, Youtube, Facebook) can complement crisis mapping from remote-sensed data, improving standard workflow of emergency mapping services as Copernicus Emergency Management Service, that provides crisis maps, actually mainly based on satellite images. Several organisations were interviewed as a part of this activity: Civil Protection Department (Headquarter and Sardegna Regional Office), National Authority for Civil Protection (Portugal), Civil Contingencies Secretariat (UK), Administration of the Republic of Slovenia for Civil Protection and Disaster Relief (Slovenia) and Doncaster Metropolitan Borough Council (UK). One of the main user requirements of the Copernicus EMS is to receive first crisis information within the first 24 h after the disaster, while today it is not unusual to experience delays up to 72 h, mainly due to the availability of the first usable post-event satellite image, caused by satellite tasking and orbital constraints, bad weather conditions preventing the collection of optical images, late activation. Moreover, the crisis maps, purely based on satellite information, have known quality limitations due to the physical constraints of satellite acquisitions (e.g. resolution, analysis technique) that affect the thematic accuracy of the analysis. The aim of this DP was to investigate the possibility of using crowdsourced social data to improve the quality and timeliness of emergency flood mapping services. During the project, crowdsourced social data (e.g. Twitter, YouTube, Pinterest), related to the historic flood event that occurred on February 2014 in the Bridgwater area, in UK, was collected. The data was analysed to generate, before the availability of a post-event satellite image, a "warning map" providing a preliminary information about the areas more affected, and to refine the quality of flood extent delineation of the final Crisis map produced once the satellite image is available.

DP4 centered on land use in the Bacchiglione river catchment area used data from citizens, social media and the last version of CORINE land cover data (CLC 2012) [23] to validate the land cover information via crowdsourced data. This activity involved stakeholders from Alto Adriatico Water Authority (AAWA), Agenzia Regionale per la Prevenzione e Protezione Ambientale del Veneto (ARPAV), Project Unit of Civil Engineers (Veneto Region), Regional Department for Soil Protection (Veneto Region), Planning Strategic Section and Cartography (Veneto Region), Urban planning department of Vicenza and Regional Forest Service of Padua. This DP involved validation of land use map CORINE land cover (CLC 2012 data set) through crowdsourcing observations by using participatory crowdsourcing mechanisms through the involvement of professional groups. The Water authority of Alto Adriatico was directly involved in the project and organised a dedicated campaign. Similar to the DP1, a professional group represented by AAWA was provided with an Android and iOS mobile phone application. The campaign was in the city of Vicenza and surroundings, directly involving volunteers while opportunistic crowdsourcing observations were also collected manually from Panoramio. Images provided by users were initially checked via a tagging API, which automatically classifies a text tag along with a confidence value. Users selected from a list of ten of the most relevant tags, which were further compared with the CLC 2012 data set values.

The range of different demonstration projects served to help understand how different types of crowdsourcing can be used to collect data to improve OS products. DP2 employed opportunistic sensing to collect vehicle mobility traces from telematics data. DP1 and DP4 employed participatory sensing in two different settings. DP3, on the other hand, employed opportunistic sensing via social media.

5 Results

This section discusses the results of the demonstration projects, and based on the experience of the project highlights various aspects need to be considered while engaging with volunteers and participants. DP1 although highly advertised via social media and association channels and shared among the hiking community and hiking associations did not provide any information via crowdsourcing. The DP followed several recommendations typically applicable for crowdsourcing such as recruiting participants through groups of special interests, incorporating the CS mechanism within their own framework as well as close communication with crowdsourcers and hiking groups to co-design the crowdsourcing apps. Several reasons could be attributed to the lack of data in the setting of the DP1—it could be possible that the hikers had a higher expectation from the information received on mobile phones to serve information beyond their practice of activities. While it is important to provide information typically unavailable to users, it is important to note that users should not be overloaded with too much information to process. Furthermore, hikers rely on well-existing practices and sources of information—there may be strong barriers for new actors that need to demonstrate long-term relevance and validity. Hikers could also be engaged and preoccupied during their hiking activities, therefore unwilling to focus on using technologies.

DP2, using passive opportunistic sensing, collected large volumes of data in the region of interest—this is primarily due to the data being collected passively without the need for users to engage in the project regularly. DP2 developed a new algorithm for detecting elevation data by combining ground-based survey data, LIDAR data and the crowdsourced mobility traces to better understand road surface elevation and subsequently improve models for estimating road pollution by accounting for the slope of each road segment. The stakeholders in the DP identified clear advantages from better understanding of EO data, which is largely unused in the sector. Another strong potential domain was identified as traffic management, where there is a need for understanding macro-regions. The approach of using crowdsourced data was seen to be strongly positive among stakeholders; however, there is a need to evaluate similar approaches on a wider region to better evaluate CORINE data.

DP3 sought to demonstrate the value of information such as images, video and text data related to crisis events, shared among social channels such as Twitter, Facebook, Flickr. Such posts are sometimes geotagged via GPS positioning sensors embedded in devices such as smartphones and tablets or can be geolocated through the toponyms of point of interests contained in the textual information. In order to understand how social media can help understand events better, a historical data set of the UK Bridgwater floods (occurring during 6th and 10th February in 2014) was collected. Stakeholders in the project were provided access to the data along with a WebGIS deployment and invited to provide their feedback and suggestions, raise concerns as well as comment on their observations. The stakeholders indicated that information from social media can provide a significant advantage and potentially save a lot of time sifting through a large number of articles and websites. It is important to note that such an experiment was on a historical data set and hence, lacked the real-time urgency Emergency Responders experience during disaster events. Indeed, taking in account the quality of social data gathered in an opportunistic way in terms of lack of geotagging, redundancy of the information, etc., it is important to develop automatic mechanisms to retrieve, filter, geolocate and ranking the social data to reduce the time spent in their analysis.

DP4 involved participatory sensing in the same spirit of DP1, but applied in a different use case and setting to understand how crowdsourced observations can help validating land use maps. The engagement of participants for crowdsourcing activities was a highly successful event. Around 1200 observations were received. Observations were post-validated based on images sent together with the land use observations. Third-party tools for image classification were used to help validate observations. A major drawback in the DP was an accessibility issue—many areas were private property, and observations about such areas were sent from locations close to the areas and hence would introduce potentially unwanted noise. Furthermore, analysis conducted in the DP highlighted that often, the resolutions of the EO data sets did not often reflect what would be in situ data—for example, the minimum size in the CLC 2012 data set for a land cover area is 25 ha and 100 m in linear scale. Smaller areas are not reflected in the data set—hence, areas such as a small vineyard surrounded by fields would be classified as 'land principally occupied by agriculture'—discrepancies such as these can be easily filled-in by in situ sensing.

While participatory sensing in DP1 generated no crowdsourced observations, DP4 was highly successful. However, part of the success could be attributed to the crowdsourcing activity being restricted to a day event. The Alto Adriatico region benefits from a very strong volunteer base that is professionally trained to respond to disasters. Finally, opportunistic social sensing provided significant insights into understanding disaster areas and providing initial warning map that can be potentially exploited to indicate evolving scenarios on the ground.

6 Experience from Demonstration Projects

The four demonstration projects provided an excellent opportunity to understand various facets of crowdsourcing when applied in practice to solve societal and technical challenges. The DPs provided a lot of insight into how crowdsourcing tools can be developed and how citizens can be engaged. One of the most interesting findings from the demonstration projects was to understand the limitations involved in engaging citizens to provide large volumes of information. While the process of developing technologies for crowdsourcing and collecting data from citizens and communities is a straightforward process, the project clearly observed there are significant challenges in the process. The primary goal of the project was to understand these challenges as feasibility studies and assess how crowdsourcing and citizen science can practically add value to space products, data sets and services. The roadmapping stage conducted in the initial stages of the project highlighted several recommendations—a primary one being the need to engage with communities by feeding back information from developed technologies. Although this was a primary consideration in DP1, where hiking maps and submitted snow reports were provided to users via the mobile application, the participation in terms of submission of reports was minimal. Several possible reasons could be attributed to this observation—users may have been unwilling to rely on sources other than their traditional medium of information. Another reason could be decreased snowfall during the year may have reduced public interest and hence participation in the crowdsourcing task. A further reason could be the practicalities involved in providing observations—taking time out during hiking to take photographs and submitting text reports on an expensive smartphone while navigating difficult terrain can be a cumbersome process and could be an unfeasible task. Further work is necessary in order to understand what could be the potential reason for reduced participation, as this was out of scope for the project. At cases such as these, it could be possible to also investigate other forms of crowdsourcing such as passive opportunistic data collection like GPS traces, wearables and smartwatches to seamlessly collect data which could be used to infer snow coverage. However, it serves to demonstrate that engaging with citizens is not a trivial task and continued interest may not necessarily guarantee crowdsourced observations.

Passive opportunistic sensing, on the other hand, provided large volumes of data, which could be successfully used by the project to improve models for pollution estimation. While a significant amount of data was collected in the project, it is necessary to be aware of practical consideration when EO data is complemented by CS data. Observations from Satellites are on a very high scale and comparing with high granularity of the data provided by CS may introduce challenges. For example, in the DP2, estimating road elevation levels and hence the slopes are critical to improve pollution models. Smaller sections such as roundabouts are designed by city planners to have minimal elevation gradients. As a result, augmenting satellite data with ground-based sensors at a larger scale are more promising as the elevation gradients are much more pronounced.

Participatory sensing data involving trained volunteers providing categories of land use data based on visual observations also needs to be handled with care. In the DP4, data was provided by trained volunteers comprising of geotagged images along with manually classified categories. A large number (1200) of observations were recorded, which was a significant success, particularly from the context of crowdsourcing and citizen science projects. However, it is important to note that the task itself may introduce noise and errors at times—for example, some areas may be inaccessible either due to difficult terrain or restricted access. In such cases, automatically geotagged images could identify the observation as relevant to a different area than the one in question. The scale of EO and CS data is different—EO provides imagery from a very high level, while in situ observations can be highly detailed. As a result, EO data may generalise information to a higher level and inconsistencies may arise—e.g. a vineyard in a farming area would be identified by volunteers as a vineyard while EO could generalise the entire area as farmland. Furthermore, discrepancies can exist in the ability of humans to observe and classify data. Some CLC 2012 data set categories such as 'urban fabric', 'roads' can be fairly easily identified by users, while categories such as 'discontinuous urban fabric' are more difficult to identify.

Crowdsourced data collected from social media can be a significant source of information—however, dealing with such data requires a great deal of consideration. Social data is high in volume and constantly increasing; often duplicated, incomplete, imprecise and potentially incorrect; informal (short, unedited and conversational) and less grammatically bounded text; generally concerning the short-term zeitgeist and covering every conceivable domain. These characteristics make automating intelligence gathering task difficult and the DP3 aimed at understanding the actionable information that can be collected from social media. The applicability of the data varies widely on the use case, and as a result, different sources provide more contextually relevant information than others. For example, in flood emergency scenarios, Twitter has resulted the most relevant channel to achieve information, not only as primary source provided by the users, but also as indirect way to access to other social data contents, through the sharing of other information channels (e.g. news, institutions and public bodies providing usually more relevant information as compared to citizens sharing information). Especially, videos and images shared on YouTube and Twitter provide immense help in understanding scenarios on the ground. Anyway, in order to exploit such data sets, there needs to be strategies in place to deal with missing information, as lack of geotagging. For instance, a very small fraction of Twitter data (<1% of analysed Tweets) actually contained geolocation information, and in most cases, the Tweets were positioned outside the area of interest. Moreover, it has to be considered a mechanism to manage the redundancy of the information and to limit the effort due to the analysis of social data. On the other hand, understanding the extent of how much a piece of information has been shared can be helpful in understanding how important or critical it is.

7 Discussion and Recommendations

Rapid changes in modern communications and mobile devices have made citizen participation in science much easier and widespread (e.g. [25]). Large numbers of volunteers can be recruited over wide geographical areas to collect, submit and interpret data at low cost [26]. Such widespread data collection (potentially over extended temporal periods) would be simply infeasible without citizen volunteered information. Indeed, such geographical spread is essential to understanding the processes behind many of the most important global challenges of today: vegetation loss, climate change, natural resource management, migration patterns, etc. In addition to geographical coverage, the volume of observation data (satellite-, airborne- and land-based)—some of which can only be interpreted by humans—is constantly growing. The ability to crowdsource detailed, high-resolution annotations of such data facilitates timely scientific analysis and decision-making. Citizen science has evolved from a hobby through to serious science and is rapidly becoming a preferred approach to conducting large-scale research [27].

The advantages of crowdsourcing extend beyond traditional scientific endeavour; the ability to mobilise a large group of volunteer data analysts (e.g., performing satellite imagery annotation) can have significant humanitarian and emergency response benefits. For example, Tomnod crowdsources object and place annotation from satellite imagery to assist in major emergencies ranging from mapping a drought-stricken area of Somalia for the UNHCR to searching for the missing Malaysia Airlines plane MH370. The FP7 SPACE Project GEO-PICTURES crowdsourced geo-referenced in situ images for the purpose of improving flood assessment from Radar EO Images [28].

Each type of citizen science and crowdsourcing initiative relies on a different degree of engagement from the citizen (and hence, a different degree of engagement on behalf of the organiser—usually an NGO or research organisation). Similarly, the geographic extent of the initiative has strong implications for the suitability of particular forms of citizen science that can be employed—none is excluded at any one particular scale but organisational and governance overheads can become problematic.

Citizen science and the wider active participation of citizens in science and governance will continue to grow. The benefits are mutual. Citizens can enjoy making a positive impact at local, national or international levels in a variety of domains (e.g. environment, ecology, medical research). Professional researchers and decision-makers can tap into an unprecedented wealth of knowledge and expertise or simply leverage the size of the crowd to collect large amounts of data across large geographical areas—all much more quickly and at must lower costs compared to traditional research approaches.

At the core of opportunities to EO activities from crowdsourcing and citizen science are the calibration and validation of satellite data and products as well as passing value to existing products and services. These can be achieved using a number of different levels of citizen science project ranging from *volunteer computing* through to *participatory sensing*. The identification of new applications or disruptive products could be achieved through hackathons and crowdsourced solution contests.

However, the use of crowdsourcing and citizen science is not without its pitfalls. The two main areas which must be considered with care are *engagement* and *data quality*. These two topics are critical to the sustained success of an initiative both in terms of maintaining the amount of data collected or processed overtime as well as the *value* of the final crowdsourced information.

Based on the findings from the demonstration projects and stakeholder analysis [24] of their understanding of how the field will evolve in the next few years, a set of recommendations is proposed:

Funding Support and Embedment

Exploitation of citizen science and crowdsourcing should be a priority in future funding calls on EO in order to support existing initiatives and foster uptake in a larger scale. This will also help to drive research into various facets of crowdsourcing such as engagement, security, privacy, user experience. Attention should also be focused on supporting wider set of participants including users not proficient in technology. Endeavours by all funders are essential to support citizen engagement and science learning at all ages. Initiatives that can help citizens relate to their own interests can have a greater acceptance and engagement, thereby fostering continued collaboration with citizens. This is a highly challenging problem and much work is needed in understanding how different forms of crowdsourcing can be used in different scenarios and settings.

Understanding

A wider understanding of the benefits of crowdsourcing could inspire new strategies and initiatives. Organisations currently exploiting crowdsourcing could provide guidelines and standards on how best to collect, analyse and reuse citizen science data (including metadata, data protection, ethics and privacy). Providing easy access and means of visualisation of the data generated by contributors can provide them with an immediate observable feedback as well as provide a sense of accomplishment. Furthermore, it is important to co-design iterative solutions with user communities to ensure a continued interest and shared ownership among user communities.

Outreach and Communication

Sharing the benefits, potential and opportunities of citizen science with different communities can be an excellent avenue for connecting with the public. One of the potential avenues could be to explore introducing citizen science within school curriculum with differentiating content and communication based on target age, starting at the very first classes in school. Team-building activities based on citizen science could be used as means for developing interest in citizen-driven collaborations. Hackathons and fab laboratories could be another means of increasing general interest, while inspiring younger people and enthusiasts to explore new avenues to bring disruptive ideas to market.

Widening Participation

Citizen science, in general, is still not representative of all in society and this unequal participation, in many cases could have inherent biases based on gender, ethnicity, age and socio-economic backgrounds. While efforts are being made to increase participant diversity, there is still much work involved in engaging large numbers of participant communities. A strong societal bias in STEM engagement and a strong bias in addressing societal and community issues could potentially result in having an incomplete and inaccurate understanding of all communities in the public. While focusing on individual communities and target groups can be helpful to provide a set of highly engaged users, there is a significant risk of alienating other pool of users and communities.

The recommendations are indeed provided as ways to start designing tools and technologies to engage citizens. The field of crowdsourcing and citizen science is highly complex that requires an in-depth understanding of users and their expectations. It is important to carefully consider all such aspects in order to build innovative business models that can leverage different approaches to engage with user communities.

Acknowledgements This research was supported by the European Space Agency funded Crowd4Sat initiative. The authors would also like to sincerely thank all the stakeholders and organizations for their inputs and interesting ideas. We would also like to thank the EU FP7 funded WeSenseIt project (Grant agreement No. 308429) and EU H2020 funded Seta project (Grant agreement No. 688082).

References

1. Haklay, M. (2013). Citizen science and volunteered geographic information—Overview and typology of participation In D. Z. Sui, S. Elwood, & M. F. Goodchild (Eds.), *Crowdsourcing geographic knowledge: Volunteered Geographic Information (VGI) in theory and practice* (pp. 105–122). Berlin: Springer. https://doi.org/10.1007/978-94-007-4587-2_7.
2. Roy, H. E., et al. (2012). *Understanding citizen science and environmental monitoring*. Technical report.
3. Goodchild, M. F. (2007). Citizens as sensors: The world of volunteered geography. *GeoJournal, 69,* 211–221.
4. Elwood, S. (2008). Volunteered geographic information: Future research directions motivated by critical, participatory, and feminist GIS. *GeoJournal, 72,* 173–183.
5. Longueville, B. D., Luraschi, G., Smits, P., Peedell, S., & Groeve, T. D. (2010). Citizens as sensors for natural hazards: A VGI integration workflow. *Geomatica, 64,* 41–59.
6. Gill, A., & Bunker, D. *Crowd sourcing challenges assessment index for disaster management*. Available online: https://pdfs.semanticscholar.org/7725/5b503bdab5b82627fa3e801042ef81bbd669.pdf. Accessed May 15, 2017.
7. Fritz, S., Fonte, C. C., & See, L. (2017). The role of citizen science in earth observation. *Remote Sensing, 9*(4), 357.
8. Irwin, A. (1995). *Citizen science: A study of people, expertise and sustainable development*. Hove: Psychology Press.
9. Bonney, R., Cooper, C. B., Dickinson, J., Kelling, S., Phillips, T., Rosenberg, K. V., et al. (2009). Citizen science: A developing tool for expanding science knowledge and scientific literacy. *BioScience, 59*(11), 977–984.
10. Cohn, J. P. (2008). Citizen science: Can volunteers do real research? *BioScience, 58*(3), 192–197.
11. Dickinson, J. L., Zuckerberg, B., & Bonter, D. N. (2010). Citizen science as an ecological research tool: Challenges and benefits. *Annual Review of Ecology Evolution and Systematics, 41,* 149–172.
12. Paulos, E., Honicky, R., & Hooker, B. (2008). *Citizen science: Enabling participatory urbanism*. Urban Informatics: Community Integration and Implementation.
13. Science Communication Unit, University of the West of England, Bristol. (2013). *Science for environment policy in depth report: Environmental citizen science*. Report produced for the European Commission DG Environment, December 2013. Available at: http://ec.europa.eu/science-environment-policy.
14. Howe, J. (2006). The rise of crowdsourcing. *Wired Magazine, 14*(6), 1–4.
15. Audubon Society. (2017). *Christmas bird count*. Retrieved 13 April, 2017 from http://www.audubon.org/conservation/science/christmas-bird-count.
16. Haklay, M. (2015). *Citizen science and policy: A European perspective*. Washington, WA, USA: The Wodrow Wilson Center, Commons Lab.

17. Lintott, C. J., Schawinski, K., Slosar, A., Land, K., Bamford, S., Thomas, D., … & Murray, P. (2008). Galaxy Zoo: Morphologies derived from visual inspection of galaxies from the Sloan Digital Sky Survey. *Monthly Notices of the Royal Astronomical Society, 389*(3), 1179–1189.
18. Lang, A. S. I. D., & Rio-Ross, J. (2011). Using Amazon Mechanical Turk to transcribe historical handwritten documents. *The Code4Lib Journal, 15.*
19. Nguyen, T. B., Wang, S., Anugu, V., Rose, N., McKenna, M., Petrick, N., … & Summers, R. M. (2012). Distributed human intelligence for colonic polyp classification in computer-aided detection for CT colonography. *Radiology, 262*(3), 824–833.
20. Yetisgen-Yildiz, M., Solti, I., Xia, F., & Halgrim, S. R. (2010, June). Preliminary experience with Amazon's Mechanical Turk for annotating medical named entities. In *Proceedings of the NAACL HLT 2010 workshop on creating speech and language data with Amazon's Mechanical Turk* (pp. 180–183). Association for Computational Linguistics.
21. Cuff, D., Hansen, M., & Kang, J. (2008). Urban sensing: Out of the woods. *Communications of the ACM, 51*(3), 24–33.
22. Maisonneuve, N., Stevens, M., & Ochab, B. (2010). Participatory noise pollution monitoring using mobile phones. *Information Polity, 15*(1, 2), 51–71.
23. Copernicus Programme Land Monitoring Service. Retrieved 15 May, 2017, http://land.copernicus.eu/pan-european/corine-land-cover/clc-2012.
24. Mazumdar, S., Wrigley, S. N., & Ciravegna, F. (2017). Citizen science and crowdsourcing for earth observations: An analysis of stakeholder opinions on the present and future. *Remote Sensing, 9*(1), 87–107.
25. Paulos, E., Honicky, R. J., & Hooker, B. (2009). Citizen science: Enabling participatory urbanism. In *Handbook of research on urban informatics: The practice and promise of the real-time city* (pp. 414–436).
26. Howe, Jeff. (2006). The rise of crowdsourcing. *Wired Magazine, 14*(6), 1–4.
27. Toerpe, K. (2013). The rise of citizen science. *The Futurist, 47*(4), 25–30. http://www.wfs.org/futurist/2013-issues-futurist/julyaugust-2013-vol-47-no-4/rise-citizen-science.
28. Validating Space Observations for Flooding with Crowdsourcing In-Situ Observations by ANSUR. Technical report 30, 2012.

Can VGI and Mobile Apps Support Long-Term Ecological Research? A Test in Remote Areas of the Alps

Laura Criscuolo, Paola Carrara, Alessandro Oggioni, Alessandra Pugnetti and Massimo Antoninetti

Abstract Long-term ecological research (LTER) is performed in many countries across the globe by scientists that gather and analyse multidecadal ecological observations and data, to support understanding and management of the environment. LTER data are used to describe the state and dynamics of an ecosystem. This valuable research is in charge of experts in ecology, and it is performed in the so-called LTER networks (for Europe, see http://www.lter-europe.net), usually organized at a national level; they consist of sites covering different ecosystem typologies, each one with research and monitoring facilities. In summer 2015, researchers of the Italian LTER network (LTER-Italy) have created a set of naturalistic trails, connecting sites of LTER-Italy (http://www.lteritalia.it/cammini), aiming to transfer and share the research results with citizens, through public events and informal science communication. Along one of these trails, performed on the Alps, the organizers proposed the use of two VGI apps to collect either biological or abiotic observations (http://www.lteritalia.it/it/content/citizenscience). The apps were chosen according to crucial characteristics, i.e. offline usage, operative system independence, strong development and support community, customization of user interfaces. This paper reports on this first test, discussing lessons learned and, in particular, the impact on the practice of research communities of participative, innovative tools, often neglected in the analysis of mobile application effects.

L. Criscuolo (✉) · P. Carrara · A. Oggioni · M. Antoninetti
CNR IREA, via Bassini 15, Milan, Italy
e-mail: criscuolo.l@irea.cnr.it

P. Carrara
e-mail: carrara.p@irea.cnr.it

A. Oggioni
e-mail: oggioni.a@irea.cnr.it

M. Antoninetti
e-mail: antoninetti.m@irea.cnr.it

A. Pugnetti
CNR ISMAR, Venice, Italy
e-mail: alessandra.pugnetti@ve.ismar.cnr.it

G. Bordogna and P. Carrara (eds.), *Mobile Information Systems Leveraging Volunteered Geographic Information for Earth Observation*, Earth Systems Data and Models 4, https://doi.org/10.1007/978-3-319-70878-2_3

1 Introduction

There is a general growing demand for scientists to engage a broader audience in their activities: this mandate is coming from a variety of actors and stakeholders, such as funding agencies, scientific journals, politicians, governments and civil society as well.

In the last century, environmental science has run a path from an exclusive top-down attitude to a participated approach. This route has also been outlined in "the three eras of environmental information" [1]: in this view, during a first era (1969–1992), environmental information was created by experts for experts; in a second era (1992–2012), it was produced still by expert, but to be shared with the public, using the Web as main dissemination medium; in the third—ongoing—era (from 2012 onwards), official data have been opened and integrated with data coming from communities' activities.

Even in the ecological field, during the last decades, initiatives aimed to participative data collection have proliferated, also as a consequence of the widespread availability of mobile devices, geographic technologies, open-source libraries and the impressive growth of the Social Web. Information collected may be of any kind: from species diversity to sensor data, to geographic reference. Citizen Science (CS, see Sect. 2.2) and Volunteer Geographic Information (VGI, see Sect. 2.2) have gained great attention both as a way of tackling research questions that could not be addressed without the involvement of large numbers of data collectors and as a method of engaging the public in the scientific process with the goal of improving scientific literacy [2]. Indeed, volunteers and private individuals from the civil society can offer scientists a context for their research and can provide novel perspectives on scientific questions, including social and policy aspects.

Also within the long-term ecological research (LTER) networks (see Sect. 2.1), there are strong needs and motivations to communicate and share with a wide audience the existence and aims of the LTER activities, as well as to increase the socio-ecological impact of LTER studies and their interactions with the civil society. In summer 2015, researchers working in the Italian LTER network (LTER-Italy, www.lteritaly.it) have created a set of naturalistic trails, connecting sites of LTER-Italy ("CAMMINI LTER", http://www.lteritalia.it/cammini/), aiming at involving citizens in their research. Along one of these trails, named "Rosa… azzurro… verde! Eco-staffetta tra i siti LTER dal Monte Rosa al Lago Maggiore" (i.e. "Pink … blue… green! Eco-trial among LTER sites from Monte Rosa to Lake Maggiore"), performed in the Italian Alps, the organizers proposed the use of two VGI apps to collect either biological or abiotic observations (http://www.lteritalia.it/content/citizenscience). The apps were chosen according to crucial characteristics, i.e. offline usage, operative system independence, strong development and support community, easy customization of user interaction environments.

This paper reports on this experience, discussing lessons learned and possible future developments, with emphasis on the impact that participative, innovative tools, often neglected in the analysis of mobile application effects, might have on the practice of LTER communities.

The next section (Sect. 2) presents the context of the research, describing both LTER and public involvement initiatives in ecology; it is followed by the description and analysis of the case study (Sect. 3). Finally, some considerations on the lesson learnt and on the future developments are proposed (Sect. 4).

2 Context

In this section, we give details regarding—first—the structure and main aims of the LTER networks, with emphasis on LTER-Italy and—second—the main applications and initiatives aimed to participative data collection in ecology.

2.1 *LTER Network and Public Involvement*

Long-term ecological research (LTER) is based on gathering and analysis of multidecadal ecological observations and data to support understanding and management of the environment. Many components of ecosystems are studied, including of course both organisms and non-living components, like air, water, mineral soil and sediments. LTER presents both a question-driven research approach and a strong monitoring component. It provides essential information about how ecosystems respond to local and global change: LTER data can be used to describe the state of an ecosystem, how it may be changing, what may be driving that change, and the probability of shifting to a higher or lower ecological quality. LTER provides a sound base of data that is particularly important in our rapidly changing world. Processes like climate change, land and sea exploitation and global trade are dramatically affecting the environment. They are altering ecosystem structure and functioning, which are at the base of the ecosystem services we depend on, like provision of food and water, air quality and water quality and the aesthetic value of a landscape.

LTER is organized in networks of sites at the national, regional and the global levels. The regional network LTER-Europe is quite complex: it comprises 25 formal national networks and more than 400 LTER sites in continental Europe and Israel, as well as a few extra-European sites (e.g. Nepal, Arctic and Antarctica). Sites in the network, with their own research and monitoring facilities, represent different ecosystem domains (terrestrial, freshwater, marine and transitional waters) and work as sentinels of ecological changes at the local, regional and global levels.

LTER-Italy is a formal member of LTER-Europe and LTER-International (ILTER) since 2006. It involves many scientific institutions (e.g. the Italian National Research Council, many universities, the Italian National Forest Service, some Scientific Societies and Public Agencies). At the moment (2017), LTER-Italy consists of a group of 28 "parent sites" belonging to terrestrial, freshwater, transitional and marine ecosystems, selected according to ILTER recommendations. Many of the LTER-Italy "parent sites" are made up by more "research sites" (around 80), managed and coordinated by public research and monitoring institutions and universities. LTER "parent sites" represent the main ecosystem typologies of Italy, and they give to LTER-Italy a strong interdisciplinary feature.

At LTER-Italy sites, as in the other LTER networks, data are related to a high variety of ecological themes; they are collected by researchers or institutional operators over the long term, with appropriate and shared methodologies.

One relevant and recent activity of LTER-Italy concerns the promotion of LTER studies to a non-expert audience, through both informal communication activities and the launch of CS initiatives (a thorough description can be found in the following Sect. 2.2, together with several examples).

CS is one of the ways to make science accessible and interactive for a range of different audiences. At the same time, empowering citizens to engage in ecological science can help scientists in answering questions that would otherwise be impractical to investigate. Actually, large programs focused on data, such as the Data Observation Network for Earth (DataONE) and the National Ecological Observatory Network (NEON), do recognize the value of Citizen Science, as evidenced by its inclusion in their development. At the European level, the European Citizen Science Association (ECSA—http://ecsa.citizen-science.net) has been established, with 17 EU countries and in collaboration with American and Australian Citizen Science associations (CSA—http://citizenscienceassociation.org and ACSA—http://csna.gaiaresources.com.au/wordpress/).

By the launch and realization of CS initiatives within LTER-Italy, researchers wish to lay the foundation for the development of a network of citizen scientists, involved in activities related to the long-term study and monitoring of ecosystems and biodiversity, with particular emphasis to some specific environmental priority issues for environmental policies. This implies the development and test of methodologies, best practices and information systems for the study of ecosystems and biodiversity with the involvement of citizens, which could be sustainable in the long term.

2.2 VGI and Citizen Science for Ecology

CS is described as "the general public engagement in scientific research activities when citizens actively contribute to science either with their intellectual effort or surrounding knowledge or with their tools and resources" [3]. CS—also known as Science 2.0—is then a practice aimed at fostering the collection of information beyond the capacity of scientific researchers in a distributed, collaborative way, by calling upon the activity of volunteer citizens [4]. Several and varied CS projects are currently ongoing over the world, involving different disciplines—from palaeontology to astronomy, from biology to medicine—and not always providing a geographic footprint together with data. When data, collected by a group of volunteers, contain a significant geographic content (boundaries, coordinates, addresses or even meaningful geotags), they are more often referred to as volunteered geographic information (VGI). This term was first coined by Goodchild as a special case of the more general Web phenomenon of user-generated content [5]. Using CS and VGI for scientific purposes is nowadays a hot topic, whose interest is testified by the growing number of scientific initiatives that aim at collecting and organizing experiences of volunteers that are eager to aid a scientific project (for instance, the SciStarter project, http://scistarter.com/; the Adventurers and scientists for conservation project, http://www.adventureandscience.org; and the Science for citizens project, http://www.scienceforcitizens.net). Most CS/VGI projects rely on huge numbers of contributors [6] and benefit from the appreciation of the scientific community. Nevertheless, many researchers are critical to the reliability of volunteered data, due to the difficulty and sometimes impossibility to assess the quality of information items, though literature and practices on this topic are growing in the last period [7, 8].

Here, follows a brief survey of CS/VGI projects focused on ecology or natural and environmental sciences. The examples show a wide variability in the project characteristics and in the strategies adopted. Referring to the categorization proposed in [9], we present some CS projects, which vary in: the way in which information is created, the tasks assigned to citizens/volunteers, the need of geographic contribution and the characteristics of citizens/volunteers. Table 1—at the end of the section—summarizes the cited examples of CS/VGI in ecology by indicating their classification with respect to the named characteristics.

In VGI projects, the *way in which information is created* and provided can be different: manually or automatically, and explicitly or implicitly. In CS, however, it is difficult to find examples where participants are completely unaware of the project purpose and provide implicitly useful information. Citizen Sort (http://www.citizensort.org/), for instance, is a website created by the Syracuse University's School of Information Studies (http://www.ischool.syr.edu), which provides a series of tools and games aimed to involve players in captivating Web adventures and, at the same time, to classify various species of insects, animals and plants. It can be considered as an example of partially implicit creation of ecological information.

Plenty of ecological projects require instead a manual and explicit contribution by citizens and volunteers, who are aware of the task and objective of their actions, must have some knowledge of the scientific problem and some training. This is the case of Monarch Larva Monitoring project (http://monarchlab.org/mlmp), which collects long-term data on larval monarch populations, and of CreekWatch project (http://www.creekwatch.org), which collects data on water quality in creek and streams.

Most ecological projects require observational measurements by means of sensors, so that information creation is automatic and explicit. This often happens also in ecological CS/VGI projects. The expertise required to the citizens/volunteers varies depending on the kind of measurements and sensors, spanning from the use

Table 1 Summary of surveyed ecological CS/VGI projects

Project name	Way of information creation	Required task	Geo-need	Volunteers' characteristics
Citizen Sort	Manual, part. implicit	Gaming/identifying	No	Neophyte
Larvae monitoring, CreekWatch	Manual, explicit	Observing/identifying	Medium	From neophyte upwards
Air Casting	Automatic, explicit	Measuring	High	From interested amateur upwards
Climate Prediction	Automatic, explicit	Computer time	No	Neophyte
Zooniverse	Manual, explicit	Recognition	No	Neophyte
eBird, MIPP	Manual, explicit	Identifying	Medium	From expert amateur upwards
Buiometria partecipativa	Automatic, explicit	Measuring	Medium	From interested amateur upwards
Snowtweet, GLOBE	Manual, explicit	Measuring	High	Neophyte
Herbaria@Home, OldWeather	Manual, explicit	Transcription	No	Neophyte
Alpine Glaciology	Manual, explicit	Measuring	High	From expert amateur upwards

of GPS devices that request small training to sophisticated sensors. Air Casting, for instance, is an open-source solution that enables mastered volunteers to collect and share environmental data (mostly fine particulate matter measurements) by coupling wearable sensors with Web or mobile apps (http://aircasting.org/).

Tasks assigned to citizens/volunteers can greatly vary also in the same scientific area. Some projects ask contributors to provide only their computer time: this is the case of the Climate Prediction initiative (http://climateprediction.net/), in which volunteers offer their computer time to run a climate model to produce predictions of the Earth's climate up to 2100 and to test the accuracy of models.

More often projects call for recognition of signs, figures and patterns that is a specific human ability in which humans excel while computers do not. By example, the Zooniverse suite (https://www.zooniverse.org/) allows citizen scientists to help researchers in several tasks related to different research fields, from recognizing floating forests in images from space (https://www.floatingforests.org) to spotting warms laying eggs (https://www.wormwatchlab.org), or interpreting historical hand-drawn tables about the African rainforest life cycle (https://www.zooniverse.org/projects/khufkens/jungle-rhythms).

The majority of CS/VGI projects in the natural field require that citizens/volunteers perform an identification of a number of objects of interest, geographically distributed (such as birds and plants) and a classification of related semantics (such as the species of the observed birds), together with other ancillary information (such as the area of observation—or geographic footprint—descriptions and photographs and time of the observation). The Ornithology Laboratory of Cornell University leads a number of such projects, as the famous eBird project (http://ebird.org) that monitors bird abundance and distribution worldwide.

Other projects require that citizens/volunteers perform observational measurements by the use of instrumentations or sensors. They can be either simple quantitative measurements that do not require complex analysis or accurate measurements by sophisticated instrumentations. Sometimes, simple measurement tools can be used: for example, the Snowtweets (http://snowtweets.uwaterloo.ca/) project requests users making simple measurements of snow cover with just a ruler and does not require any interpretation of data or analysis of the type of snow being measured.

Other projects may need sensors to perform the measurement task: nowadays, personal instruments integrated into mobile devices are available for several computing usages. Through them, people can participate in collecting and sharing measurements of the everyday environment that matter to the public, allowing to make even more detailed observations of their communities. The already mentioned Air Casting provides an example of this practice.

Also, the GLOBE Student Climate Research Campaign (http://globe.gov/web/scrc/overview) and the Italian "Buiometria Partecipativa" (http://www.pibinko.org/bmp2/) projects lead student and common citizens in measuring environmental parameters, by following protocols and using low-cost devices.

In some projects, users are required to make transcription tasks: they are given existing data and are asked to transcribe it into another form, more usable by project scientists. For example, the Herbaria@Home (http://herbariaunited.org/atHome/) and OldWeather (http://www.oldweather.org/) projects scan historical paper records from over a hundred years ago and ask participants to read the information and load it into the project database. Most of this activity is a relatively simple recording of other people's work and can be performed by less-experienced or beginner citizen scientists.

Facing the *need of geographic information from contributors*, most projects about objects identification (birds, insects, plants and so on) call for a geographic reference (geographic footprint). In some of these projects, the precision and the accuracy of geographic information are not necessarily too high; one can tolerate certain levels of imprecision in geo-locating objects of interest, while a greater accuracy is needed in classifying the seen objects. This is, for example, the case of the MIPP project (http://lifemipp.eu), which aims at monitoring nine endangered species of insects.

In other projects, geographic information with high spatial accuracy is required: these are generally named geographical CS projects. It is the case, for instance, of a recent academic research regarding the glaciological information produced by aware and unaware volunteers on Italian Alps [10]: measurements recorded during hikes, boundaries drawn by OpenStreetMap (https://www.openstreetmap.org), GPS tracks, geo-located photographs of glacier fronts and other kind of non-expert contributions have been used to harvest precious information about the glaciers' variations through time.

The typology of the involved volunteers—that is, linked to their motivation and expertise—has a great impact on the projects and on the quality of created information. In order to characterize citizens/volunteers generating Web contents, we use the categories proposed by [11]. They consider five major types of volunteers, arranged in ascending order of experience and increasing complexity of their features and activities: the neophyte, the interested amateur, the expert amateur, the expert professional and the expert authority. A further type of volunteer is the unaware volunteer, whose characteristics may be heterogeneous with respect to the experience and engagement [9].

3 Volunteered Contribution to LTER Researches

LTER-Italy researchers have been recently debating on the role that CS/VGI can play in their work and particularly in the monitoring of ecosystems. Indeed, the variety and the extent of ecological components that are studied would require broad and capillary observations, which cannot be collected by a small number of scientists, nor by a limited number of sensors. With reference to the characteristics listed in the previous section, volunteers—expert or not—can contribute in LTER research by creating information manually or automatically and by playing various task: observing, identifying, measuring and recognizing. Here are some practical examples of tasks that can be assigned to different typologies of volunteers:

Expert amateurs can collect natural observations and make species identifications (observation and identification tasks);
Neophytes and amateurs can take basic measurements, for instance, temperature, water level, snow level, …, following given guidelines (measurement tasks);
Neophyte and amateur volunteers can pick photographs and videos on the LTER sites or report about the general state of the ecosystems (observation tasks); and
Expert amateurs can warn about possibly critical conditions or novelties, for instance, station damages, the presence of alien species,…(observation and identification tasks).

In general, it has been recognized that expert professionals, expert amateurs, interested amateurs and neophytes can help.

They all need portable instrumentation to collect, archive and share information in various formats (e.g. text, numerical values, images, multimedia contents) during field trips.

While the modern mobile technology provides for GPS sensors, cameras, memory storage units and network wireless connection, mobile apps supply the appropriate framework to data collection, validation and sharing. In fact, in the context of LTER sites, a suitable mobile application must respond to a series of requirements, depending on the researchers' needs, on the usability and on the user-friendliness, on the terrain and landscape conditions in order to involve and keep engaged volunteers.

These ideas should be verified, and the next section describes how LTER-Italy researchers tried to meet these requirements and how they tested some mobile applications during one of their research trails, i.e. the participative initiative performed in Summer 2015, by creating a case study.

3.1 VGI in a LTER Trial: A Case Study

Since the early months of 2015, researchers working in the Italian LTER sites laid out a participative activity that was totally original for the network. A team composed by institutes from the National Research Council of Italy (CNR), universities and other Public Research Institutions planned a set of three naturalistic trails, connecting LTER sites in different regions of Italy. The trails have been scheduled in order to be performed by heterogeneous groups of researchers, thematic experts and interested citizens, by walking and cycling together, while observing the peculiarities of the surrounding environment.

The whole event, named "Cammini LTER", has been conceived as a sort of pilgrimage of ecological research: citizens had an opportunity to familiarize with the components and conditions of Italian ecosystems, from the Mediterranean Sea to Alpine tundra, while experts and researchers involved them in ecological issues and in their fieldwork.

During the event, researchers experienced new ways to divulgate and communicate science. Moreover, in one of the trails, the scientific coordinators wished to involve citizens not only in itinerant informative activities, but also in the provision of VGI, in order to test the ideas described in Sect. 3. This route, named "Pink… blue…green! Eco-trial among LTER sites from Monte Rosa to Lake Maggiore", has been planned to connect three LTER-Italy research sites: one of the high-altitude Alpine sites ("Mosso Institute") on the Monte Rosa, the high-altitude lakes "Laghi Paione" and the subalpine lake "Lago Maggiore". It consisted of six legs, to be covered mainly by foot, walking across some Alpine valleys. The schedule provided a kick-off on 22 August at Gressoney-La-Trinité (Aosta); then, the route moved its first steps on 23 August from the LTER site "Istituto scientifico A. Mosso" of the University of Turin on the Monte Rosa. It reached on 27 August the site of Alpine lakes "Lago Paione Inferiore" and "Lago Paione Superiore" to attain on 28 August the final destination: the site "Lago Maggiore", one of the large Italian subalpine lake, managed by the Istituto per lo Studio degli Ecosistemi (Institute for Ecosystem Study, ISE—CNR) in Verbania. The variety, also cultural, of environments and landscapes along the route, was great. Populations living in lake areas or in the ancient alpine villages (among which the Walser minority), witness ages of fruitful alliance between man and nature and at least one century of environmental researches.

The planned steps and stops are illustrated in Fig. 1.

At the Alpine site "Mosso", on Monte Rosa, where Alpine tundra dominates, the main research activity concerns the study of the characteristics of soils, waters and snow, with a particular focus on the effects of the environmental factors (e.g. snow cover duration, soil temperature and moisture, air temperature) on the carbon and nitrogen dynamics and on vegetation characteristics.

The Alpine lakes (Laghi Paione) are permanent oligotrophic environments: their fragility and ecological vulnerability make them early-warning indicators of climate change. They are studied mainly for assessing the long-term effects of atmospheric pollutants (acidifying compounds, heavy metals, nitrogen) and climate change on the hydrochemical and biological features of the lakes (phytoplankton, zooplankton and benthos taxonomic composition).

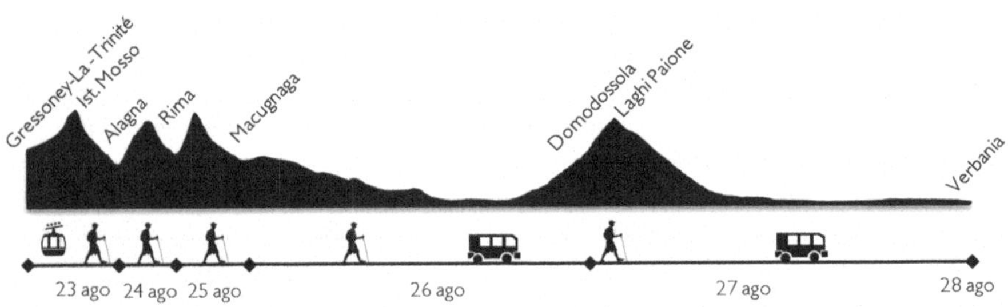

Fig. 1 Graphical schedule of the "Pink…blue…green!" trail

Lago Maggiore is the second largest and deepest Italian subalpine lake. It has been studied since the beginning of the last century, for evaluating the trophic state evolution of the lake and, more recently, the effect of global warming. Limnological research covers chemistry, hydrology, physics, plankton ecology, microbial ecology and palaeolimnology.

Decisions about the citizens' participation to the trail had to consider its natural features, the characteristics of the sites visited (e.g. their technological facilities), the activities usually performed by scientists and their requests with respect to CS/VGI. As previously described, the involved LTER-Italy sites were very different, in terms of both environmental characteristics and research performed; so many issues needed to be answered: Which kind of tasks should be entrusted to non-expert participants? Which volunteer characteristics could be expected? Which the best device for field information provision? In summary, which experimental design would best fit the researchers' and the volunteers' needs?

After some consultations and a strong debate, researchers drafted a list of tasks to be performed by the volunteered participants during the trail legs, going from identification of species to measurement of physical parameters and geomorphological observations. They are summarized in Table 2, together with the ecological characteristics and the traditional research activities led by LTER scientists.

3.2 The Solution Adopted: VGI Apps for LTER Trials

To support volunteers in performing the tasks of Table 2, the solution proposed consisted one or more mobile applications. During the first test edition, due to time and resource limits, existing—customizable—applications have been adopted, instead of building brand new ones. A list of requirements has been compiled, and a screening has been done on the available applications, in order to select those that best fulfil the list.

Some of the requirements enlisted respond to technological and scientific needs: they are aimed to provide a strong framework for data collection and a suitable structure to volunteers' contribution. These scientific requirements are, for instance, the possibility for scientific coordinators to administrate and validate the collected contributions; the possibility to get (or convert) data in standard formats and use it on different platforms; the possibility to perform a quality check; and the possibility to customize the app interface and functionalities. Besides, a suitable licensing and policy, a robust technology and an active support community are among the technological needs.

Some other requirements indeed have been considered in order to meet the challenges related to the field usage and to encourage the participation of volunteers. These "usability requirements" comprehend the compatibility of the app with most mobile devices and operative systems; the offline usage (most LTER sites are not covered by the Internet connection); a user-friendly interface; and some

Table 2 Summary of the ecological characteristics and the activities led—by volunteers and LTER researchers—for each trail leg

Trail leg	Environment	Volunteers tasks	LTER activities
1	High mountain, Alpine tundra	Nivological observations (snow presence, melting status, photographs); Soil observations (colour, pH, temperature); Rocks observations and identification; Air physical measurements (temperature and humidity)	Physical and chemical features of the soil and the snow Effect of snow covering on the biogeochemical cycles Plant phenology
2	High mountain, Alpine tundra	Air physical measurements (temperature and humidity); Species identification (flora and fauna)	Physical and chemical features of the soil and the snow; Effect of snow covering on the biogeochemical cycles; Plant phenology
3	High mountain, Alpine tundra	Air physical measurements (temperature and humidity); Species identification (flora and fauna)	
4	Alpine forest	Air physical measurements (temperature and humidity); Species identification (flora and fauna)	
5	Alpine lake, internal water	Air physical measurements (temperature and humidity); Species identification and sampling (benthos and plankton); Hydrological observations and measurements	Zoobenthos; Phytoplankton; Zooplankton; Nutrient cycles; Biogeochemical processes
6	Large subalpine lake	Air physical measurements (temperature and humidity); Species identification and sampling (plankton); Hydrological observations and measurements	Phytoplankton, zooplankton and benthos taxonomic composition; Population dynamics of the main species

mechanisms for users' interaction and social reward. All requirements appear in the first column of Table 3.

Two applications resulted to better respond to the requirement list. The first of them, iNaturalist (iNat), is dedicated to species identification and is complimentary to a contributive Web platform (http://www.inaturalist.org/), maintained by the California Academy of Sciences and powered by open-source software. Although it is focused on the sole species identification purpose, it has strong mechanisms for users' interaction and for contributory quality control. The software provides also for useful customization and administration tools.

Table 3 Response of the selected applications to basic and supplementary requirements

Requirements	iNat	EpiCollect
Basic requirements		
Robust technology	++	++
Offline usage	++	++
Compatibility with most devices	++	+
Data access (viewing and management)	++	+
Control on data quality	++	+
Customization	++	++
Free and open source	++	++
Supplementary requirements		
Active support community	++	+
Mobile and Web	++	+
Easy to use	++	++
Interaction among users	++	
Users support	+	
Multipurpose		++

The second application selected is EpiCollect (http://www.epicollect.net/) [12], and it has been developed at the Imperial College London. It is a multipurpose, Web and mobile app, which supports data collection by providing customized forms (questionnaires) and hosting archives of geo-referenced contributions. Its usefulness rests on the ability to adapt to various research topics involved in the LTER activities, in particular for the abiotic and geomorphological parameters.

Table 3 shows how the two applications respond to the mentioned requirements that have been divided in basic (unavoidable) and supplementary ones.

Table 4 shows how the volunteers' tasks, distributed along the trail legs, take advantages of the selected apps.

The applications underwent a customization to enable the volunteers to report the expected information, within the specifically configured project "LTER-Italy". The customization regarded the project settings (geographical and temporal extension, description, graphical settings,…) and the configuration of the Web

Table 4 Distribution of volunteered tasks among the trail legs and the selected mobile applications

Volunteers tasks	Trail legs	iNat	EpiCollect
Nivological observations	1		X
Soil observations	1		X
Rock observations and identification	1		X
Air physical measurements	1, 2, 3, 4, 5		X
Species identification	2, 3, 4, 5	X	
Hydrological observations and measurements	5		X

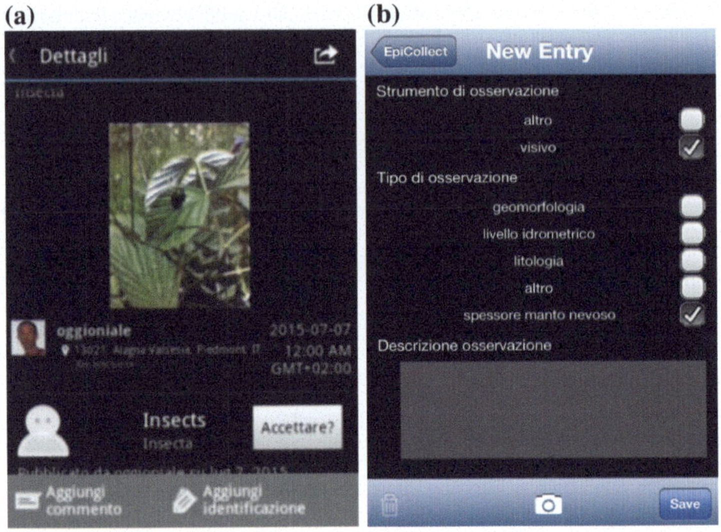

Fig. 2 Screenshot from the iNat application, LTER-Italy project (**a**) and from the EpiCollect application, LTER-Italy customized (**b**)

forms, addressed to reporting information. A special attention has been paid in the configuration of the EpiCollect forms, in order to get a VGI archive compliant with the OGC Observation and Measurement (O&M) standard (http://www.opengeospatial.org/standards/om). To achieve this goal, any field in the Web form has been conceived to create a database entry, which corresponds to an element in the O&M schema. In this way, the VGI items collected by participants via the mobile app are assimilated to standard sensor observations and—as such—can be delivered (and accessed) through dedicated, interoperable Web services (OGC Sensor Observation Service).

The mobile apps configured have been installed on several devices, running with different operative systems. LTER-Italy researchers tested their functioning on field before the event begins. In Fig. 2, some screenshots from the applications can be seen.

In order to assist participants in installing the apps, defining the project settings and reporting their contributions, some brochures and tutorials have been released (some tutorials are available at http://www.lteritalia.it/content/citizenscience).

4 Discussion of Results and Future Developments

Eventually, the trail has been performed according to the schedule. More than 50 researchers and a variable number of volunteers (depending on the leg) took part in the trail, among which many thematic experts from Italian and European institutions. They played the first actual use in the field of LTER-Italy applications for data collection and have reported their strengths and weaknesses. They collect 147 VGI contributions and identified 89 different species (see Fig. 3). The results of their activities are documented on the LTER-Italy website (http://www.lteritalia.it/content/osservazioni-dai-cittadini).

The CS/VGI experiment has had both positive outcome and unexpected difficulties: both offered lessons that we discuss here following.

In general, we can affirm that the LTER trail event produced multiple results, both material (the trail itself, videos, photographs, presentations for the public, observations collected) and non-material (personal experience of researchers and the public, ideas on this new research approach, the impact on local communities, etc.). We are here more interested in investigating if and how the innovative tools offered to collect VGI in the case study of this trial show an impact on the participants and on the LTER research.

From a technological point of view, the two apps did not present critical issues and were easily used by the volunteers during the trail.

They satisfied all the requirements (see Table 3) and proved to be fairly reliable in the Alpine environment, where connections are rare.

It seems that the main obstacle in their usage was a lack of motivation of the volunteers: their adoption should be more fostered by a clear mandate to the participant volunteers, who should be openly charged by researchers in the data collection task. To this purpose, training sessions and engaging activities should be prepared in advance and become an intrinsic part of the whole process.

However, this mandate deeply depends on the trust of researchers in the volunteers' participation. In fact, in this first experience, researchers working on the LTER sites, involved by the trail coordinators, proved to be often sceptical with respect to the advantage of volunteers' involvement and seldom able to openly and fruitfully express cases in which they could be supported by more or less expert public.

Fig. 3 Map view of the collected observations within the LTER-Italy project on iNat

Researchers were ready to interpret "*Citizen Science*" like "*Science to inform Citizens*", not like "*Science made with Citizens*".

Therefore, this first test has shown that also researchers need time and approaches to include CS and VGI in their everyday practices and in formulating scientific questions that can be answered by/with citizens and volunteers.

The more, LTER-Italy researchers are rather unprepared to conceive other ways to obtain the public support besides data collection, i.e. co-creation of scientific questions, data elaboration, result check and analysis, as described in the "VGI and Citizen Science for Ecology" section.

This experience proved that many CS activities could be performed in the future in LTER-Italy. However, they require including the individuation and the implementation of tools able to foster citizens/volunteers participation. At the same time, also the researcher's trust in the non-researcher's ability to contribute in ecological research must be leveraged, so that that concepts of CS could really become a part of the scientific activities in LTER, from the cultural and experimental point of view.

Citizens and researchers should start working together with every stage of the research process, from the creation of the scientific questions to the elaboration of the results and of their impacts on society and on the environmental policies. This process should include also the phase of creation and configuration of the mobile applications for LTER studies.

The lessons learned from the trail "Pink…Blue… and Green" will be relevant for the future of LTER-Italy and for the other trails that are planned in the next future. Actually, the "Cammini LTER" has become a real "brand" for LTER, and it will continue and develop also in the next years. This initiative is also spreading at the international level, within the LTER-International community, where a new project called "TRAIL" has started, with a global perspective. "TRAIL" aims at stimulating CS within the ILTER network, in different cultural, economic, scientific and political contexts. It will provide a unique opportunity to examine various facets of CS, with a comparative approach, enabling to explore the scientific, pedagogical and ethical dimensions of CS in different geographical contexts.

Acknowledgements This work has been partially funded by the Italian Flagship Project RITMARE, by LifeWatch-Italy (the Italian component of LifeWatch ERIC) and by the LTER-Italy network.

References

1. Muki Haklay talk in the Wilson Center in Washington DC on 'Environmental Information—The Roles of Experts and the Public', 29 April 2014.
2. Miller-Rushing, A., Primack, R., & Bonney, R. (2012). The history of public participation in ecological research. *Frontiers in Ecology and the Environment, 10*(6), 285–290.
3. Socientize Consortium. (2014). White paper on citizen science for Europe.
4. Hand, E. (2010). Citizen science: People power. *Nature, 466*(7307), 685–687.

5. Goodchild, M. F. (2007). Citizens as voluntary sensors: Spatial data infrastructure in the world of Web 2.0. *International Journal of Spatial Data Infrastructures Research, 2,* 24–32.
6. Kelling, S., Hochachka, W. M., Fink, D., Riedewald, M., Caruana, R., Ballard, G., et al. (2009). Data-intensive science: A new paradigm for biodiversity studies. *BioScience, 59,* 613–620.
7. Dorn, H., Törnros, T., & Zipf, A. (2015). Quality evaluation of VGI using authoritative data —A comparison with land use data in Southern Germany. *ISPRS International Journal of Geo-Information, 4,* 1657–1671.
8. Antoniou, V., & Skopeliti, A. (2015). Measures and indicators of VGI quality: An overview. *ISPRS Annals of Photogrammetry, Remote Sensing & Spatial Information, 1,* 345–351.
9. Bordogna, G., Carrara, P., Criscuolo, L., Pepe, M., & Rampini, A. (2014). On predicting and improving the quality of Volunteer Geographic Information projects. *International Journal of Digital Earth, 9*(2), 134–155. https://doi.org/10.1080/17538947.2014.976774.
10. Criscuolo, L., Pepe, M., Seppi, R., Bordogna, G., Carrara, P., & Zucca, F. (2013). *Alpine* Glaciology: An historical collaboration between volunteers and scientists and the challenge presented by an integrated approach. *ISPRS International Journal of Geo-Information, 2*(3), 680–703. https://doi.org/10.3390/ijgi2030680.
11. Coleman, D., Georgiadou, Y., & Labonte, J. (2009). Volunteered geographic information: The nature and motivation of produsers. *IJSDIR, 4*(1), 332–358.
12. Aanensen, D. M., Huntley, D. M., Feil, E. J., al-Own, F., & Spratt, B. G. (2009). EpiCollect: Linking smartphones to web applications for epidemiology, ecology and community data collection. *PLoS ONE, 4*(9), e6968. https://doi.org/10.1371/journal.pone.0006968.

Part II
Methods and Techniques for VGI Creation, Management and Analytics

Toward Citizen-Edited Image-Populated Ontologies for Earth Observation—A Position Paper

Robert Laurini and Imed Riadh Farah

Abstract Ontologies are now very popular to organize knowledge categories in various domains. Ontologies can be defined as semantic networks based on concepts. In geography, several ontologies were designed essentially organizing landforms and ecosystems. But all the ontology concepts are textual. In remote sensing and aerial photos, it is important to deal with images, i.e., leaf concepts are no more textual, but multimedia. The scope of this paper will be first to give some hints in order to design image-populated ontologies. Indeed, let us consider, for instance, a glacier: depending on the kind of channel used for remote sensing, its corresponding images will look differently. The idea will be to design an ontology in which each landform will be accompanied by several sampled images with their relative metadata so to give a sort of visual ontology. Those samples will be the starting point for adapted image-processing algorithms. In addition, in some cases, some places such as fords or polluted zones will appear very differently; moreover, in the case of caves, they will not appear at all. So, the second idea will be not to ask experts but to ask citizens located at the proximity to identify those places therefore enriching the visual ontology.

Keywords Geographic knowledge · Remote sensing · Ontologies
Visual ontologies · Multimedia ontologies

1 Introduction

In geography, ontologies are more and more common. They are used to organize knowledge, to make some automatic reasoning, and to construct systems for interoperating several geographic databases. In essence, an ontology is a kind of

R. Laurini (✉)
INSA-Lyon, University of Lyon, Lyon, France
e-mail: Robert.Laurini@gmail.com

I. R. Farah
La Manouba University, Tunis, Tunisia

G. Bordogna and P. Carrara (eds.), *Mobile Information Systems Leveraging Volunteered Geographic Information for Earth Observation*, Earth Systems Data and Models 4, https://doi.org/10.1007/978-3-319-70878-2_4

semantic network in which concepts and terms are organized into classes which have often three types of relations between them, namely "is_a," "has_a," and "part_whole." Some ontologies regrouping geographic features have been built only with those relations, but more and more spatial relations are included into geographic ontologies. However, in those ontologies, features are described textually [19].

In Earth observation, the starting point is a collection of images issued from remote sensing. Those images are taken from different satellites with different technologies and characteristics using different channels. The result is that the same geographic feature, e.g., a lake, appears very differently.

As a consequence, the idea is to design a visual ontology. In other words, it will be a graph of textual concepts, but the bottom leaves will be samples of remote sensing images. The second aspect of this ontology is that it will be populated by any citizen. The scope of this paper will be to analyze ontologies for Earth observation and explain how they can be constructed with sampled images with the help of citizens.

This paper will be organized as follows. First, a state of the art for both ontologies and ontologies for Earth observation will be developed. Then our proposed approach for a visual ontology will be explained. Some perspectives will conclude this paper.

2 Ontologies and Ontologies for Earth Observation

The scope of this section will be to give some details regarding ontologies, geographic ontologies, and ontologies for Earth observation.

2.1 Ontologies and Geographic Ontologies

In this section, we will successively present the key elements of ontologies together with the common methodologies to construct them.

2.1.1 Components of Ontologies

Nowadays, several computer applications use ontologies as a model of a domain. This use allows us, in particular, to solve various problems such as the design and indexing of databases, integration and sharing of data, the semantic web, and also geographic information retrieval and territorial intelligence [17, 18].

An ontology may take a variety of forms, but necessarily it will include a vocabulary of terms and specifications of their meanings [29]. It includes definitions and an indication of how the concepts are connected, which collectively impose a structure field and constrain the possible interpretations of the terms.

Maedche and Staab [21] say that an ontology encompasses a glossary of concepts, relations, and axioms. With reference to various studies [11, 28], we can summarize the primary components of an ontology as follows:

- The "concepts," also called terms or classes of ontology, corresponding to relevant abstractions of the problem domain, selected according to the objectives we give ourselves and to the intended use of the ontology.
- The "relationships" are types of interactions between classes (or concepts) in an ontology. In the geographic context, topological relationships define meaningful relationships because of the spatiotemporal dependence of geographic features [17, 18].
- The "axioms" are used to model the conditions that are always true for the domain.
- The "instances" are the extensional definition of ontology. They represent unique elements conveying knowledge (static, factual) about the problem domain. For instance, the Eiffel Tower in Paris is an instance of the class Tower.

2.1.2 Approaches to the Development of an Ontology

Since their emergence, ontologies represent a solution to many problems such as data integration, software interoperability, data sharing, knowledge engineering. There are many methods in engineering ontology, but the lack of structured and common guidance slows the development of ontologies within and between teams, the extension of any ontology, and its reusability. We mean by the term methodology, work procedures and the steps that describe why and how of the conceptualization and then the built artifact.

There are five approaches to the ontology design: inspirational, deductive, synthetic, collaborative, and inductive [13]. Let us present them very rapidly.

A. **The inspirational approach**

In the inspirational approach, a designer takes decisions alone to gather the terms of the domain analysis, design, and verification of ontology. The developer must be both domain expert and expert in ontology design to ensure the success of the design, and above all, to ensure the adoption of the ontology by the user community. This process depends heavily on the creativity of one person, his/her inspiration, and his/her personal perception of the area.

B. **The deductive approach**

With a deductive approach, the general principles are first adopted and then processed and applied to the target domain. The resulting ontology can be seen as an instance object of these general concepts.

C. **The synthetic approach**

In the synthetic approach, a set of related ontologies is identified. The developer then synthesizes the elements of these ontologies with the concepts of the new target area, producing a newly unified ontology.

D. **The collaborative approach**

The mark of a "modern" ontology is its large size and high complexity. This kind of ontology encompasses a rich set of knowledge that its understanding (full) exceeds that of any single developer or designer or even a small team of designers. The development of a large-scale ontology must be the fruit of a joint effort of several domain experts and software designers.

E. **The inductive approach**

With the inductive approach, ontology is developed by observing, examining, and analyzing a specific case or cases in the domain of interest. The characterization of the resulting ontology for a specific case is applied to other cases in the same field. The design is based largely on the widespread cases selected during development.

2.1.3 Geographic Ontologies

A geographic ontology is an ontology whose concepts have a spatial dimension. We define a geographic ontology as a set of geographic concepts that describe the geographic space or entities or phenomena of this geographic area [18] claims the necessity of integrating spatial and overall topological relationships. The components of a spatial ontology are the same as those of a non-spatial ontology. However, they have some characteristics in their definitions. The majority of the concepts of these ontologies represent an object or a real phenomenon localized in space at a given time, while relationships can be topological between objects; we mention for example inclusion, adjacency, equality, etc.

Geographic ontologies can be classified into three categories [6]:

- Ontologies of space are specifically dedicated to the description of concepts that characterize space as point, line, etc. These ontologies are typically developed by leading bodies of standardization.
- Ontologies of geographic area model the concepts of hydraulic data, an ontology with concepts of urban data, or an ontology describing the concepts of electrical data networks, etc. These are "trade" ontologies developed by the user community in the field.
- The spatial ontologies (or spatiotemporal), which are ontologies whose concepts are located in space. A temporal component is often needed as an addition to the modeling of geographic information because geographic applications also often wield temporal or spatiotemporal data.

Geographic ontologies have specific needs [6] that are related to the necessity to define spatiality by using various types of spatial data (line, point, or surface…), types of space objects (i.e., objects with spatial attributes), spatial relations as topological relationships, and/or continuous fields (raster) intensionally define spatial concepts by using axioms containing spatial predicates and reason on the spatiality of instances objects, i.e., inferred from the spatial relationships describing the set of valid spatial relationships.

Several projects have been set up to develop ontologies; we can mention:

- Towntology project [2]. The authors aimed at defining an ontology for urban planning. It provides domain experts with a framework for assistance in information retrieval, indexing their documentation, and staff training.
- An ontology for risk and disaster [24]. The authors have proposed a conceptual ontology which allows to describe the knowledge of risk and disaster domain. The target is to identify the key concepts to characterize events, risks, accidents, disaster, and the related concepts, and then, to organize these concepts together with their relationships. They provide a useful framework for analyzing different types of events, whether located or diffused, of natural, industrial, or social origin.
- Ordnance Survey [12]. The researchers of this organization have created several ontologies on the study of the hydrological characteristics whose interest was to improve the use of data for their clients and facilitate the semi-automatic data processing. They created these ontologies with the help of experts and field workers. Their approach was to form from these data a glossary including all the domain concepts and then to make ontologies with it.
- SUMO [23]. This project is different from others because it consists in developing a high-level ontology called Suggested Upper Merged Ontology (SUMO) in the field of geography and transports. The advantage of this approach is the reuse of the general concepts of this high-level ontology for the creation of other ontologies.
- MIDA [30]. This project comes within the framework of the improvement of an atlas of coasts. The researchers created an ontology by atlas. Their approach is distinguished by determining all domain concepts according to 5 categories, such as discipline, theme, or place.

2.2 *Ontologies for Earth Observation*

In Earth observation, remote sensing provides a useful source of data that may be used in several environmental applications such as meteorology simulations and environmental protection. In order to better interpret a remotely sensed image content, researchers switched from a pixel-based image analysis paradigm to an object-based image analysis, since scene components such as houses and trees are the only meaningful entities in an image, which are of interest for users [3]. To cope with the phenomenal increase of RS images flood, automatic processing, and annotation of such data are necessary. Ontologies have been considered as a good support for knowledge representation and reasoning, thus enabling the automatic annotation of RS images content by modeling the properties of observable objects as well as their spatial relationships. Actually, several geographic-related ontologies have been proposed in the literature

In this section, we present some previous works on ontology-based image annotation.

The Fodomust ontology presented in [8] attempts to represent urban objects in satellite images. This ontology is composed of 91 concepts, 20 attributes (in total), and 66 leaf concepts. The depth of the ontological tree is 6. See Fig. 1.

Each concept of this ontology is characterized by contextual, spatial, and spectral attributes which are enriched by a priori values.

The ontology of the DAFOE platform [4] containing concepts associated with data stemming from the European database of land biophysical occupation "Corine Land Cover," enriched by other concepts and semantic relations. It covers agricultural areas, land use, water areas, etc. In addition, it offers three families of spatial relationships between its concepts, namely topological, distance, and directional relations.

For instance, *SatellitesSceneOntology* is a well-known satellite ontology (Fig. 2). This ontology presents many levels: upper-level or domain ontologies. Upper-level ontologies, such as DOLCE, OBOE, and SWEET [26, 27], hold the general core classes of a given domain to interface both top and domain ontologies. Domain ontologies include specific classes to describe the vocabulary related to a specific domain by specializing the terms introduced in the upper-level ontologies.

By analyzing the cited state of the art, we can deduce these comments:

- Some ontologies regrouping geographic features have been built only with those relations, but more and more spatial relations are included into geographic ontologies. However, in those ontologies, features are always described textually.

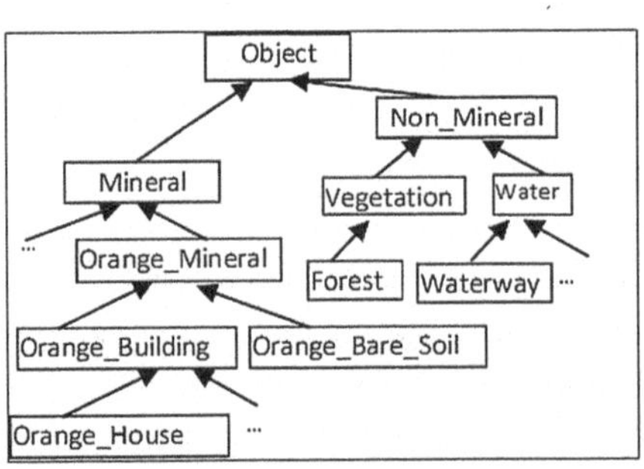

Class	Attribute name	Values
spectral	spectral_signature_Blue	[21.7-62.3]
	spectral_signature_Green	[19.4-80.1]
	spectral_signature_Red	[29.7-135.1]
	spectral_signature_NearInfaRed	[34.8-139]
	spectral_signature_SBI	[14.6-60.1]
	spectral_signature_NDVI	[50.2-108]
spatial	diameter(m)	[13-61]
	area (m^2)	[10-600]
	perimeter (m)	[28-116]
	elongation (m)	[1-3.1]
	Miller index	[0.5-0.8]
	Solidity index	[0.85-1]

Fig. 1 Excerpts of the Fodomust ontology [8]

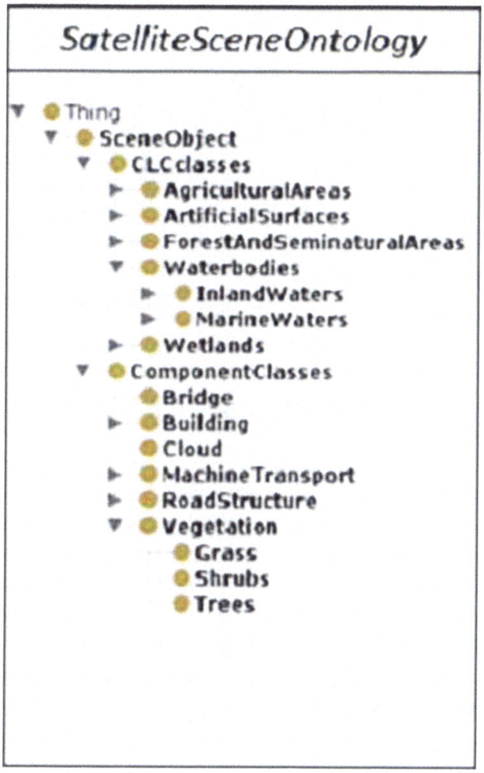

Fig. 2 Satellite scene ontology

- The existing ontological resources do not represent the sensor of acquisition, and especially, the temporal characteristics of remote sensing images; however, those elements can appear as metadata (or data about data).
- The remote sensing images are not used as information in the ontology development process.
- The majority of the existing ontologies are not rich in axioms and in rules which represent knowledge of the domain. Remember that axioms and rules make mechanisms of inferential reasoning in the ontology.

Now that key elements of Earth observation ontologies are given, let us examine what volunteered information can bring.

3 Volunteered Geographic Information and Ontologies

The term VGI was introduced by Goodchild [10] as the widespread participation of a large number of citizens, who often have little official skills, to the creation of geographic information. Indeed, VGI is the set of data collected and produced:

(i) by using geo-mobile devices, (ii) by annotating the geographic features using geo-web mapping interfaces, and (iii) by the extraction or deduction of location information from geospatial data in social media: photos, videos, blogs, tweets, etc. [5]. These various sources can be described as voluntary sources where the data are generated and provided voluntarily by users [15].

Crowd-sourced data in the form of Volunteered Geographic Information (VGI) have emerged as a new promising source of information presenting multiple benefits for business and local authorities. This new source of information presents multiple advantages compared to traditional expensive ways of geographic information acquisition such as satellite remote sensing techniques.

3.1 Ontologies for VGI Incorporation

Among the various advantages of VGI compared to traditional geographic information source, there is firstly the price of acquisition of geographic data which have always been owned by specialized companies and accessible, as proprietary contents are today completely effortless and priceless thanks to VGI. Another advantage is that VGI offers real-time data which are extremely needed in case of critical applications such as emergency services or disaster management decision making [16].

In the past decades, disaster management has been successfully backed by Spatial Data Infrastructures (SDIs) to provide reliable geospatial information making sharing of data between organizations an easier task [16]. However, SDIs suffers from lack of real-time data that can lead to inaccurate and delayed decisions in serious situations. Interesting approaches have emerged to resolve this problem by incorporating VGI in SDIs so that they can offer more accurate and real-time data. However, VGI data are mostly unstructured, heterogeneous and come from diverse sources (such as social networks) using different languages, vocabularies, and concepts that may describe the same phenomena with different terms. This highlights the need for a common vocabulary to eliminate misunderstandings that can lead to severe consequences especially in critical applications such as disaster management. Recently, the professional way of dealing with conceptual diversity is the use of ontologies that can perfectly respond to this problem as explained earlier. In this case, ontologies can be used as a support to semantically filter VGI data in order to be incorporated into SDIs.

3.2 Enriching Ontologies with VGI Data

Another interesting way to exploit VGI data is to develop domain ontologies based on crowd-sourced information. The advantages of this are to identify the recent and improvable spatial data by means of a common understanding, i.e., the concepts and terminologies of disaster information shared over the social media such as Twitter and Facebook [16]. By this way, ontologies can be more evolutionary providing current and reliable information needed to understand an actual situation before deciding the actions.

3.3 Integrating VGI and Authoritative Data Using Ontologies

Integrating multi-source geographic information is an interesting research subject in order to improve data reusability and facilitate exchange of increasing amounts of data and supporting the increasing need of geographic information in different applications and fields [1, 9].

The emergence of VGI has recently heightened the need of integrating geographic data and combining official with VGI data in order to improve the freshness of specialized companies' datasets [25].

Integrating two or more geospatial datasets involves the need of mapping heterogeneous objects of different datasets at geometric and semantic levels and can be possible only if the different datasets to be integrated shares a unique common knowledge which is not always possible. This leads us to the major problem of semantic heterogeneity in which the same conceptualizations can be referred with the same concepts and this is even worse when using VGI data. In this context, ontologies can be a very adequate solution since ontologies offer the possibility of making a common vocabulary [25] have proposed a method to resolve such problem by using mappings from datasets to the domain ontology which results as new datasets semantically interoperable that can be used for quality comparison purposes, or queried for more specific results. See also [20] and [7].

3.4 Potentialities of VGI

VGI presents an interesting potential for ontologies. In fact, ontologies can be powered by generated content from social networks (Twitter, Facebook, Linked In…) and discussion forums.

These contents are an important resource for the exploitation of ontologies. But also they can be enriched in order to present an informative result based on user interactions. For this, it would create ontologies associated with this VGI that would connect citizens at the international levels.

Thus, VGI can contribute to the development of ontologies in three main ways:

- *Content:* Analysis and interpretation of indicators and calculation methodologies may be enriched by critics and observations of experts to the structures concerned by the ontologies in the discussion forums. The comments are stored in totality and filtering (validation) would be made by competent persons and experts (persons responsible for ontologies) just to save those deemed relevant. The selected comments may be a potential source for interpretations.

- *Network organization:* A social network for the ontologies will strengthen cooperation and exchanges between users by creating a permanently active "Ontologies community."
- *Dissemination of data:* Outputs (reports…) can be disseminated to the general public through their sharing in social networks.

4 Proposed Approach

In this study, we propose to design image-populated ontologies for remote sensing imagery. For this, each landform will be accompanied by several sampled images with their relative metadata to give a visual ontology. Those samples will be the starting point for adapted image-processing algorithms.

Let us mention two considerations: (i) in some places, fords or polluted zones will appear very differently, (ii) in the case of caves, they will not appear at all. So the second idea will be not to ask experts but to ask citizens residing at the proximity to identify those places, therefore, enriching the visual ontology.

4.1 Ontology Development Process

The proposed ontology is based on the reuse of existing ontological resources, the learning from the remote sensing images, and the formalization of the domain knowledge (citizens, landform, natural risks, knowledge on scenes, sensors of acquisition, etc.). The development process is composed of three steps as shown in Fig. 3.

Step 1 *Choose of the core ontology*

The first step in the process is to choose the core ontology which represents basic knowledge about remote sensing image domain. According to the literature review made on the existing geographic ontologies, we propose to reuse the Fodomust ontology presented by [8] and to consider it as core ontology. Indeed, this ontology represents the spatial and the spectral specificities of objects detected in a satellite image.

Step 2 *Evolution of the core ontology*

This step attempts to update the selected core ontology to represent spectral and temporal dimensions.

Step 3 *Enrichment of the core ontology*

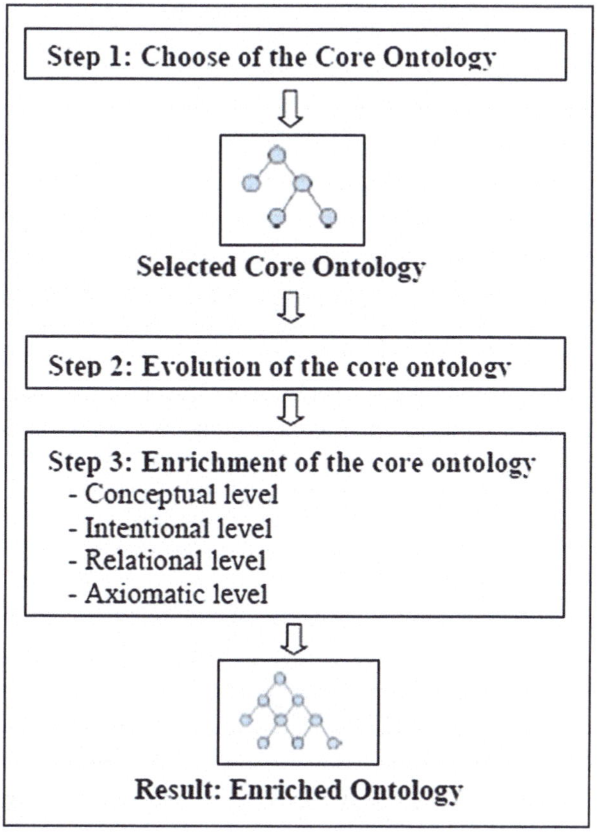

Fig. 3 Proposed ontology development process

The enrichment process is established at conceptual, intensional, relational, and axiomatic levels.

1. *Conceptual enrichment:*

The conceptual enrichment is based on the reusing of existing ontological resources in order to enrich the core ontology by new concepts. We adopted a process of enrichment composed of the steps as presented in [22]:

(a) *Discovery of candidates:* identification of existing ontological resources to be integrated into the core ontology.
(b) *Evaluation:* determination of reusable concepts from each ontological resource. This step is based on the mapping between the core ontology to be enriched and each candidate resource. The alignment is made at the linguistic and structural level.

(c) *Integration and placement:* placement of the selected concepts (from the previous step) in the core ontology based on the extracted mappings between concepts.
(d) *Quality and coherence check:* check the consistency of the ontology during its enrichment.

2. *Intensional enrichment:*

In this enrichment level, we try to enrich ontology concepts by a priori values from remote sensing images. This enrichment allows guiding the semantic annotation of remote sensing images. Figure 4 shows the process of this enrichment.

3. *Relational enrichment:*

The spatial relations between objects allow maximizing semantics of the satellite scene. Therefore, we propose to reuse the fuzzy ontology of spatial relations proposed in [14]. In fact, this ontology is enriched with a fuzzy representation representing topological, distance, and directional relations between objects, based on operators of the mathematical morphology to extract fuzziness zones of every object, and on fuzzy operators (*t*-standard, *t*-conorms, etc.) to establish the spatial relationships.

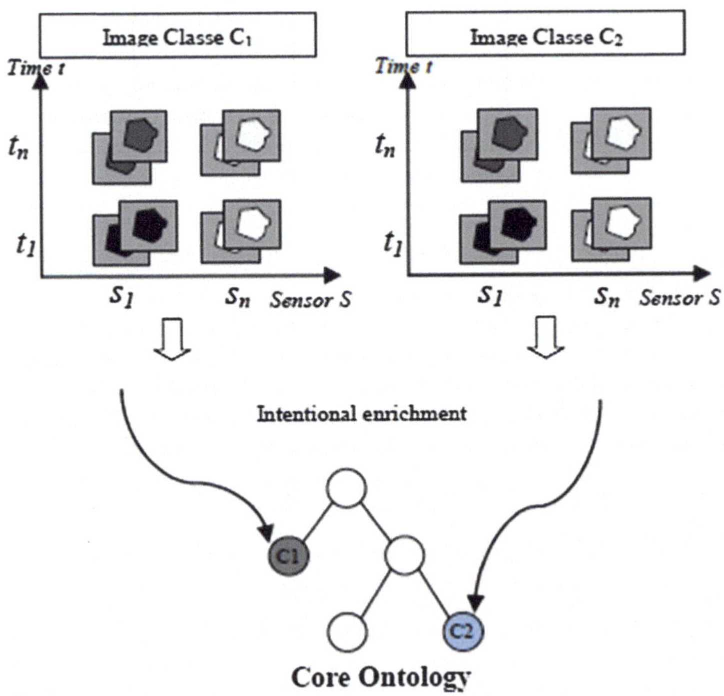

Fig. 4 Intensional enrichment of the core ontology

4. *Axiomatic enrichment:*

This level allows enriching the ontology by axioms and rules to maximize its ability on reasoning. Thus, we propose to add a set of axioms which represents knowledge related to the natural risks and phenomena such as forest fire, erosion, flood. Next, we present some rules as follows:

R1 *If the agricultural ground is neighboring a mountain having a slope* >10°, *Then erosion risk is Strong.*

R2 *If the agricultural ground is neighboring a mountain having a slope* <2°, *Then erosion risk is Low.*

4.2 Formalism of the Proposed Visual Ontology

Our visual ontology represents three levels of the remote sensing image interpretation:

- Pixel level: through the intentions of the concepts;
- Region level: through the concepts; and
- Decision level: through axioms and rules of domain;

The visual ontology (VO) is formalized as follows:

$$\mathrm{VO} = \left\{C, P, A, \leq^{C}, \mathrm{Tags}\right\}$$

in which:

- C is the set of concepts of domain
- $P = \{Pf | pf(c, \alpha, \pi)\}$

  where

- $c \in C, \alpha$ is the class of the attribute: geometric, spectral.
- π is the weight attributed to the attribute.

- $\leq^{C}: C \times C$ is a hierarchical relation between concepts.
- A is a set of axioms defined in the fuzzy language.
- Tags is the set of information produced using social networks.

4.3 Ontology-Based Image VGI

According to the proposed visual ontology, this approach allows annotating remote sensing image by meta-information describing its semantic content with VGI data. The process is described as follows:

Step 1 *Region extraction*

(a) Extraction of regions from the request remote sensing image using clustering algorithms such as fuzzy *C*-Means, region growing.
(b) Extraction of features related to each region such as radiometry values, texture, shape.
(c) Computing the spatial relationships between extracted regions.

Step 2 *Region classification*

In this step, we propose a fuzzy-matching process between the extracted regions and the concepts of the ontology. The goal is to find the best concept which describes the requested region.

This process is described as follows:

(a) We compute the similarity index $S(R_i, C_i)$ between a region

$R = \{Feature_1, Feature_2, \ldots\}$ of the image and the concepts
$C = \{Attribute_1, Attribute_2, \ldots\}$ of the ontology representing the attributes of concept C

(b) The concept which has the higher membership degree is considered as a class for the request region.

Step 3 *Generating annotation*

In this step, we apply the fuzzy-matching process for all regions in order to generate the global annotation of the requested image. The generated annotation has the following structure.

```
<Visual Ontology>
      <Characteristics>
            <Sensor>… </Sensor>
            <Bands> …</Bands>
            <Resolution> …</Resolution>
            <Geo-Ref>…</Geo-Ref>
            <Url> …</Url>
      </Characteristics>

      <Objects>
            <Object>
                  <id_objet>… </id_objet>
                  <Concept_ontology>…</ Concept_ontology
                >
                  <Features>
                        <Radiometry>… <Radiometry>
                        <Texture >…</Texture>
                        <Form>…</Form>
                  </Features>
            </Objet>
      </Objects>

      <Tags>
            <Citizens Comments>
                <Citizen Comments>
                      <Social Network>
                              <Object>…</ Object>
                            <Image>…</ Image>
                      </Social Network>
                </Citizen Comments>
            </Citizens Comments>

            <Experts Comments>
                <Local Comments>
                        <Object>…</ Object>
                </Local Comments>
                <Global Comments>
                        <Image>…</ Image>
                </Global Comments>
            </Experts Comments>
      </Tags>

</Visual Ontology>
```

5 Experimental Study

In Nefzi et al. [22], an experimental study was carried out on conceptual enrichment by reusing existing ontological resources. Initially, the selected core ontology, i.e., Fodomust [8], contained 91 concepts. Table 1 summarizes the result of enrichment steps, respectively, at conceptual, intensional, relational, and axiomatic levels.

Table 1 Experimental results of the enrichment process

Enrichment level	Enrichment result
Conceptual level	Adding new concepts – 34 concepts from ontology of hydrology – 65 concepts from FTT-01 ontology
Intensional level	Creating of base of 60 images class: – Multi-dates: January and July – Multi-sensors: Spot and LandSat Generating of 60 Gaussian fuzzy membership functions representing 60 classes.
Relational level	Adding of 15 fuzzy spatial relations
Axiomatic level	Formalization of 100 rules modeling the natural risk of erosion

In this section, we present an experimental application showing the visual ontology-based image annotation process (Fig. 5):

(a) Let's consider an image representing a scene acquired by a SPOT sensor in March 2000 (Fig. 5a).
(b) Extraction of regions: supervised segmentation using ISODATA algorithm (Fig. 5b).
(c) Extraction of features of regions using ENVI tool (Table 2).
(d) Generation of the visual ontology in XML format.

Here is a part of generated visual ontology in XML format.

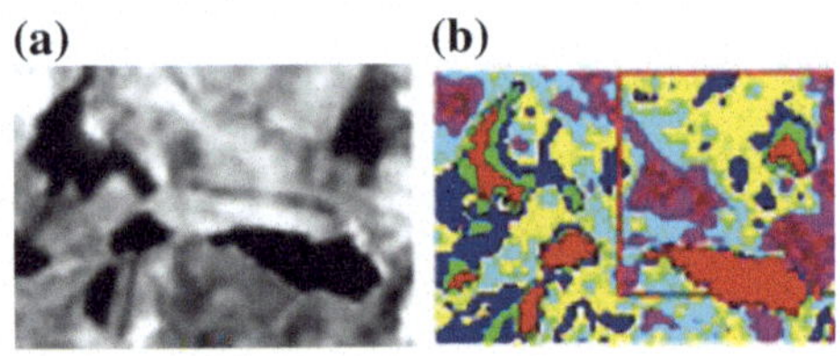

Fig. 5 **a** SPOT image acquired in March 2000; **b** extracted regions from the image

Table 2 Results of the matching

Region	Fuzzy-Matching	Returned concept
A	S(A, lake)=0.8314 S(A, forest)=0.3279 S(A, ground) = 0.01958 ...	Lake, Score = 0.8314
B	S(B, Wheatfield)=0.74565 S(B, forest)=0.4987 ...	Wheatfield, Score = 0.74565
C	S(B, ground)=0.6980 S(B, forest)=0.2153 ...	Ground, 0.6980

```
<Visual Ontology>
        <Characteristics>
                <Sensor> SPOT </Sensor>
                <Bands> 3 </Bands>
                <Resolution> 20 meters</Resolution>
                <Geo-Ref>
                        <Proj> UTM, Zone 32N </Proj>
                        <Datum> Carthage </Datum>
                </Geo-Ref>
                <Url> "http://www.isa2m.rnu/carthage.dat" </Url>
        </Characteristics >

        <Objects>
                <Object>
                        <id_objet>A</id_objet>
                        <Concept_ontology>
                                <Name_concept> Lake </Name_concept>
                                <MembershipDegree>            0.8314
                             </MembershipDegree>
                        </ Concept_ontology >
                        <Features>
                                <Radiometry>
                                <band name = "XS1"> 59 </band name>
                                <band name = "XS2"> 41 </band name>
                                <band name = "XS3"> 16 </band name>
                                <Radiometry>
                                <Texture >
                                        <Haralick                 fea-
                                     ture>..</Haralick feature>
                             </Texture>
                                <Form> … </Form>
                        </Features>
                </Object>
                …
             <Url>        "http://www.is2m.rnu/CarthageObjects.dat"
     </Url>
        </Objects>
        <Tags>
          <Citizens Comments>
              <Citizen1 Comments>
                   <Social Network>
                     <Object>
                             <Type> A </ Type>
                             <MembershipDegree>
                     0.7856</MembershipDegree>
                      </ Object>
                     …
                     <Image>
                             <Type>Urban Planning </ Type>
                             <MembershipDegree>
                     0.8841</MembershipDegree>
                     </ Image>
                 </ Social Network >
```

```
            </Citizen1 Comments>
                    ...
          </ Citizens Comments>

         <Experts Comments>
             <Local Comments>
                    <Object>
                           <Type> A </ Type>
                             <MembershipDegree>
                           0.7856</MembershipDegree>
                     </Object>
                    ....
             </Local Comments>
             <Global Comments>
                    <Image>
                           <Type>Urban Planning </ Type>
                             <MembershipDegree>
                           0.8841</MembershipDegree>
                      </Image>
             </Global Comments>
           </Experts Comments>
           ...
         </Tags>
  </Visual Ontology>
```

6 Conclusion

In this chapter, we have presented an image-populated and citizen-edited ontology approach for remote sensing image annotation.

By doing so, we have defined a new type of ontology, i.e., visual ontology in which leaf concepts are no more textual but images. Moreover, this research can be extended in one hand to visual taxonomies and on the other hand to multimedia ontologies, for instance, to store sounds or noises in cities.

This visual ontology was based on the reuse of existing ontological resources, and knowledge of the domain. In fact, the enrichment process is established at several levels such as conceptual, intensional, relational, and axiomatic level. The main goal of this visual ontology is to reduce the semantic gap between low-level image features and their semantics. Thus, we proposed a matching process between regions of an image and concepts of the ontology. The goal is to find the best concept which describes image regions. This process is simulated on images representing a natural scene.

References

1. Al-Bakri, M., & Fairbairn, D. (2012). User generated content and formal data sources for integrating geospatial data. *International Journal of Geographical Information Science, 26* (8), 1437–1456.
2. Berdier C., & Roussey C. (2006). Urban ontologies: The towntology prototype towards cases studies. In J. Teller J. Lee C. Roussey (Eds.), *Ontologies for Urban Development: Interfacing Urban Information Systems*, (pp. 143–156). Germany: University of Geneva, Springer Verlag.
3. Blaschke T., Hay G. J., Kelly M., Lang S., Hofmann P., Addink E. et al. (2014). Geographic object-based image analysis–towards a new paradigm, *ISPRS Journal of Photogrammetry and Remote Sensing 87*, 180–191.
4. Charlet, J., Szulman, S., Aussenac-Gilles, N., Nazarenko, A., Hernandez, N., Nadah, N., et al. (2010). *DAFOE: une plateforme pour construire des ontologies à partir de textes et de thésaurus" 10ième Conférence Internationale Francophone sur l'Extraction et la Gestion des Connaissances*. Tunisie: Hammamet.
5. Cinnamon, J., & Schuurman, N. (2013). Confronting the data-divide in a time of spatial turns and volunteered geographic information. *GeoJournal, 78*(4), 657–674.
6. Cullot N., Parent C., Spaccapietra S., & Vangenot C. (2003). Des SIG aux ontologies géographiques. *Revue internationale de Géomatique, 13*(3) 285–306. Lavoisier.
7. Derbal K., Bordogna G., Pasi G., & Alimazighi Z. (2015). Spatial querying supported by domain and user ontologies: An approach for web GIS applications. FQAS, volume 400 of *Advances in Intelligent Systems and Computing*, (pp. 353–365). Springer.
8. Durand N., Derivaux S., Forestier G., Wemmert C., Gançarski P., Boussaid O. et al. (2007). Ontology-based object recognition for remote sensing image interpretation. In 19th *IEEE International Conference on Tools with Artificial Intelligence*, 2007, pp. 472–479.
9. Fonseca, F. T., Egenhofer, M., Agouris, P., & Câmara, G. (2002). Using ontologies for integrated geographic information systems. *Transactions in GIS, 6*(3), 231–257.
10. Goodchild, M. F. (2007). Citizens as sensors: The world of volunteered geography. *GeoJournal, 69*(4), 211–221.
11. Gruber, T. R. (1993). *Toward principles for the design of ontologies used for knowledge sharing*. Padova, Italy: International Workshop on Formal Ontology.
12. Hart G., Dolbear C., & Goodwin J. (2007). Lege Feliciter: Using structured english to represent a topographic hydrology ontology.
13. Holsapple Clyde W. & Joshi K. D. (2005). A Collaborative approach to ontology design. *Communications of the ACM, 45*(2).
14. Hudelot, C., Atif, J., & Bloch, I. (2006). *Ontologie de relations spatiales floues pour l'interprétation d'images, Rencontres francophones sur la Logique Floue et ses Applications, LFA-2006*. France: Toulouse.
15. Kitchin, R. (2013). Big data and human geography: Opportunities challenges and risks. *Dialogues in Human Geography, 3*(3), 262–267.
16. Koswatte S., McDougall K., & Liu X. (2014). Ontology driven VGI Filtering to Empower next Generation SDIs for Disaster Management.
17. Laurini R. (2015). Fundamentals of geographic engineering for territorial intelligence. In A. Perez Gama (Ed.), *Knowledge Engineering Principles, Methods and Applications*, (pp. 1–56). New-York: Nova Science Publishing.
18. Laurini R. (2017). *Geographic knowledge infrastructure: applications for territorial intelligence and smart cities*. Elsevier, 310 p.
19. Laurini R., & Kazar O. (2016). Geographic ontologies: Survey and challenges. *Journal for Theoretical Cartography, 9* 1–13. ISSN 1868-1387.
20. Le B. T., Dieng-Kuntz R., & Gandon F. (2004). On ontology matching problems for building a corporate semantic web in a multi-communities organization. In 6th *International Conference on Enterprise Information Systems*, pp. 236–243.

21. Maedche, A., & Staab, S. (2001). Ontology learning for the semantic Web. *IEEE Intelligent Systems, 16,* 72–79.
22. Nefzi, H., Farah, M., Farah, I. R., & Solaiman, B. (2014). *A semi-automatic mapping selection in the ontology alignment process, KEOD'2014* (pp. 459–466). Rome: Italy, October.
23. Pease A., Niles I., & Li J. (2002). The suggested upper merged ontology: A large ontology for the semantic web and its applications.
24. Provitolo D., Dubos-Paillard E., & Muller J. (2009). Vers une ontologie des risques et des catastrophes: le modèle conceptuel. *Ontologie et dynamique des systèmes complexes, perspectives interdisciplinaires.*
25. Ramos, J. M., Vandecasteele, A., & Devillers, R. (2013). Semantic integration of authoritative and volunteered geographic information (VGI) using ontologies. In *Association of Geographic Information Laboratories for Europe* (AGILE).
26. Raskin, R. G., & Pan, M. J. (2005). Knowledge representation in the semantic web for earth and environmental terminology (sweet). *Computers & Geosciences, 31*(9), 1119–1125.
27. Tripathi, A., & Babaie, H. A. (2008). Developing a modular hydrogeology ontology by extending the sweet upper-level ontologies. *Computers & Geosciences, 34*(9), 1022–1033.
28. Uschold M. (1996). Building ontologies: Towards unified methodology, Presented at the 16th Annual conference of the *British Computer Society Specialist Group on Expert Systems*, Cambridge, UK.
29. Uschold, M., Healy, M., Williamson, K., Clark, P., & Woods, S. (1998). Ontology reuse and application. In N. Guarino (Ed.), *Formal Ontology in Information Systems (FOIS'98).* Trento, Italy: IOS Press.
30. Wright, Dawn J., Dwyer, N., & Cummins, V. (2010). *Coastal Informatics: Web Atlas Design and Implementation.* USA: Information Science Reference.

A Network of Low-Cost Air Quality Sensors and Its Use for Mapping Urban Air Quality

Philipp Schneider, Nuria Castell, Franck R. Dauge, Matthias Vogt, William A. Lahoz and Alena Bartonova

Abstract Recent rapid technological advances in sensor technology have resulted in a wide variety of small and low-cost microsensors with significant potential for measuring air pollutants. In this contribution, we evaluate the performance of a commercially available low-cost sensor platform for air quality and show how the data from a network of such devices can be used for high-resolution mapping of urban air quality. Our results indicate that the sensor platforms are subject to a significant sensor-to-sensor variability as well as strong dependencies on environmental conditions. A field calibration of all individual sensor devices by co-locating them with an air quality monitoring station equipped with reference instrumentation is thus required for obtaining the best possible results. We further demonstrate that, despite relatively low accuracy at the individual sensor level, a methodology based on geostatistical data fusion is capable of merging the information from the sensor network with model information in such a way that we can obtain realistic and frequently updated maps of urban air quality. We show that exploiting the "swarm knowledge" of the entire network of sensors is capable of extracting useful information from the data even though individual sensors are subject to significant uncertainty.

1 Introduction

Poor air quality continues to pose a major health risk in many regions worldwide. The World Health Organization reports that approximately 3 million deaths related to air pollution occur each year globally [48]. In Europe alone, close to 500,000 people die prematurely each year due to causes attributable to air pollution [12]. Air quality is generally observed in accordance with legislative requirements by monitoring stations that are equipped with state-of-the-art reference equipment. While they do provide very reliable and highly accurate observations for a wide

P. Schneider (✉) · N. Castell · F. R. Dauge · M. Vogt · W. A. Lahoz · A. Bartonova
NILU—Norwegian Institute for Air Research, PO Box 100, 2027 Kjeller, Norway
e-mail: ps@nilu.no

G. Bordogna and P. Carrara (eds.), *Mobile Information Systems Leveraging Volunteered Geographic Information for Earth Observation*, Earth Systems Data and Models 4, https://doi.org/10.1007/978-3-319-70878-2_5

variety of air pollutants, such stations tend to be large, complex, require significant maintenance by skilled personnel, and are therefore very expensive to set up and operate. Due to this reason, there are generally only very few of such sites available in the urban environment. While this scarcity of monitoring data is already problematic in European and North American cities, it is particularly dramatic for developing countries. In such locations, air pollution is often severe, but no sufficient funding is available for deploying adequate monitoring networks. Furthermore, the official monitoring sites are often strategically placed at highly polluted roadside locations in order to allow checking for compliance with the regulation. While such locations are, therefore, useful for policy compliance, significantly fewer such monitoring stations tend to be available in areas where the majority of the population lives.

Recent advances in sensor technology have enabled the construction of small and low-cost platforms for measuring various parameters related to air quality [1, 5, 32, 37, 43, 44]. It should be noted that in this context the concept of "low cost" should be interpreted relative to the cost of traditional air quality monitoring using stations equipped with reference instrumentation. The prices of low-cost platforms at this point typically range from a few hundred to a few thousand of Euros. The devices are designed to be almost maintenance-free and can be used by operators with relatively limited technical skills. The cost of the traditional technology they are compared to, however, involves both high initial purchasing cost and high maintenance expenses. A fully equipped reference measuring site requires a shelter including numerous analyzers equipped with an adequate sampling system, certified calibration standards, and a data acquisition system taking care of data logging and transmission. All this equipment must be kept at a stable temperature, which requires an air conditioning device. Overall, such a station is typically worth around 100 k Euros, but this only represents the most obvious part of the costs. In addition, all analyzers must be regularly calibrated and serviced by highly skilled personnel, both on-site and at dedicated facilities. It is challenging to quantify these maintenance costs, given that they are dependent on numerous parameters such as the involved equipment, the size of the network, and the region. However, the annual cost of operating such a station is generally equivalent to a relatively large fraction of the total station value.

Low-cost platforms for measuring air quality are ideally suited to be used within a crowdsourcing or citizen science framework. Due to their small size and lower cost, such devices can be deployed throughout the urban environment at much higher density than what is feasible with traditional air quality monitoring stations equipped with reference instruments.

Figure 1 shows as an example a comparison of the spatial distribution of air quality monitoring stations and a recently deployed network of low-cost microsensors. It should be noted that the number of air quality monitoring stations in Oslo exceeds the legal requirements by the EU air quality directive, and typically a city of the same size would have even fewer stations. A large and dense network of low-cost sensors is thus capable of providing significantly more spatial detail than traditional reference stations. Therefore, such networks are likely able to

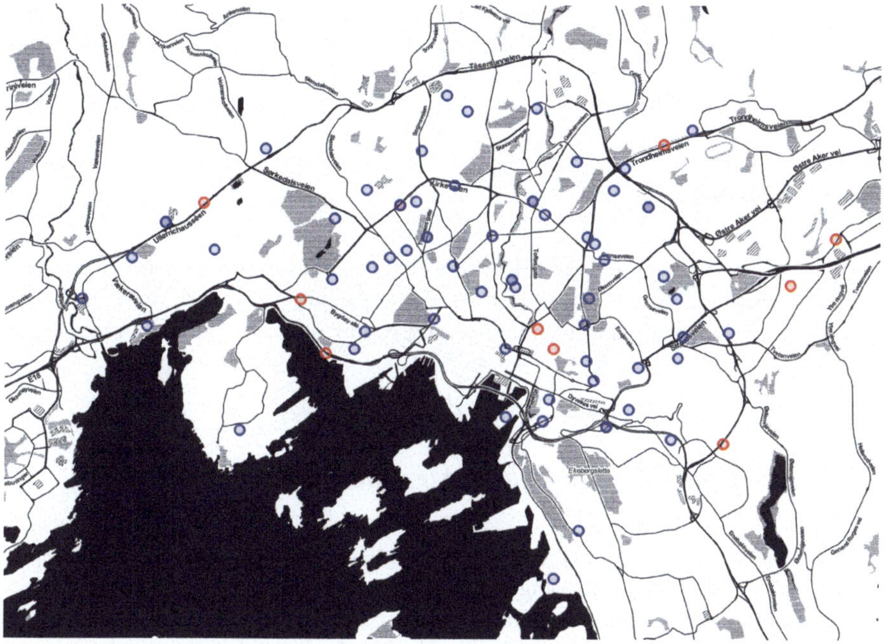

Fig. 1 Comparison of the typical deployment density and geographical coverage of a network of official air quality monitoring stations (red markers) versus a network of low-cost microsensors (blue markers), here shown for Oslo, Norway. It is obvious from this comparison that a network of deployed microsensors is able to provide significantly more spatial detail about urban air quality. Background map data provided by the OpenStreetMap project, tile design by Stamen Design

better characterize the distribution of air pollutants in the environment and thus have the potential for providing complementary information to air quality monitoring stations.

Despite the increased deployment density, such sensor networks exhibit still large data gaps in space and, to a lesser extent, in time due to sensor failures, etc. In contrast, many relevant applications require spatially exhaustive maps of air quality throughout the urban environment. In addition, the sensors suffer from significantly higher uncertainties than traditional monitoring equipment [2, 6]. Due to these reasons, it is challenging to use the information from the sensors alone for mapping purposes. In order to overcome these limitations, we present here a method for combining the observations from a network of low-cost air quality sensors with time-invariant data from a local-scale air pollution dispersion model.

In this contribution, we first summarize the general background with regard to air-quality-related crowdsourcing and citizen science efforts using low-cost microsensors as well as previous work on combining information obtained from such activities with model information for air quality mapping purposes. Subsequently, we present some of our recent work that was carried out in Oslo, Norway. We first present results from evaluating the performance of a commercial low-cost sensor platform. We then present a methodology for exploiting the information acquired by an entire network of such sensors for mapping urban air quality.

2 Background

In the following section, we provide a short overview of existing air-quality-related crowdsourcing and citizen science efforts, describe previous work on data assimilation as well as data fusion techniques, and report on some previous research on using these methods for applications in urban-scale air quality mapping.

2.1 Crowdsourcing for Urban Air Quality

In recent years, technological progress has resulted in a paradigm shift with respect to low-cost sensors for measuring various air pollutants such as nitrogen dioxide (NO_2), nitrogen oxide (NO), carbon monoxide (CO), ozone (O_3), black carbon, and particulate matter (PM_{10}, $PM_{2.5}$) [2, 28, 31, 44]. Integrated within a platform for communication through the mobile network or directly linking to a user-carried smartphone, such devices are generally inexpensive, small, and lightweight. While their accuracy is at this point not sufficient to replace observations from calibrated reference air quality instrumentation [2, 6, 44], their other characteristics mean that they can be deployed in large numbers throughout a region: for example within a citizen science or crowdsourcing framework. Such a high-density network of low-cost devices has the potential to provide spatial and temporal information about urban air quality at spatial scales and resolutions that were not possible in the past with conventional monitoring systems.

Several studies have investigated the potential use of this new generation of air-quality-monitoring devices, particularly for mobile measurements using portable sensors. Hasenfratz et al. [13] presented a system for using small and portable devices for monitoring ozone concentration using smartphones and showed how to perform an on-the-fly calibration of low-cost gas sensors based on co-located measurements. The potential of crowdsourced observations of air quality to be used in conjunction with statistical land-use regression models has also been demonstrated for mapping urban-scale air quality [14]. Mead et al. [32] deployed a network of low cost, electrochemical gas sensors in Cambridge, UK, and found that even though these devices were designed for sensing at parts-per-million mixing ratios, they can actually be used at the parts-per-billion level given suitable configuration and operation. They conclude that sparse, static, traditional monitoring stations are not capable of fully characterizing the urban environment and that low-cost microsensors are able to provide a much more complete assessment of the small-scale spatial and temporal variability of urban air quality. Peters et al. [35]

describe the use of bicycle-based mobile measurements for measuring PM_{10} and ultrafine particles along a street network and found significant temporal variability. Similarly, Peters et al. [36] demonstrate the bicycle-based measurement of black carbon and ultrafine particles for studying the micro-variability of street-level air pollution and the exposure of bicyclists. Castell et al. [5] show how mobile technologies and low-cost sensors can be applied for environmental monitoring and in particular for measuring air quality at the street level. Other studies use novel technologies such as smartphones and personal air pollution sensors to examine the diurnal variability in personal air pollution levels and the relationship between modeled and measured exposure to air pollutants in different microenvironments [3, 34] used bicycle-based mobile monitoring to map black carbon in an urban environment and found that it can provide insight into the spatial variability within a city, given a sufficient number of passes along each road segment in order to reduce short-term temporal variability. Moltchanov et al. [33] deployed a network of metal-oxide wireless sensors and found that in general the individual network nodes exhibited high inter-node consistency and sensitivity to their respective local microenvironments. However, a sensor-specific temporal variation of the calibration parameters was observed, which was corrected using observations from a nearby air quality monitoring station.

2.2 Data Assimilation and Data Fusion for Urban Scale Mapping of Air Quality

Data assimilation, of which data fusion is a subset, in general, is a way of combining model information with observations in a mathematically objective way (by providing the best linear unbiased estimate). As such it provides the possibility to create self-consistent and realistic representations of the Earth system [24, 29, 30]. In this process, value is added to both the observations and the model: The spatial and temporal gaps in the observations are interpolated in a meaningful way, while the model is constrained by the observations.

Data assimilation is carried out using many different techniques. Some data assimilation methods can actively interact with the model. These include variation methods such as 2D-VAR, 3D-VAR, and 4D-VAR [30], and sequential methods such as the Kalman Filter (KF) [23], the extended Kalman Filter [4] and the ensemble Kalman Filter (EnKF) [10]. Lahoz and Schneider [30] provide a comprehensive overview of these types of data assimilation methods. Other data assimilation methods, which we refer to here as data fusion techniques, are based on a wide variety of different statistical techniques including regression and geostatistical methods [7, 11, 17, 21, 47]. The goal of all these methods is to combine the various data sources and to provide an optimal estimate of the spatial distribution of the parameter in question [8]. While spatial interpolation techniques have been used in the past for data fusion purposes in various fields, not many studies exist that apply such methods for urban-scale air quality mapping.

Van De Kassteele et al. [26] successfully performed statistical mapping of annual surface PM_{10} concentrations over Western Europe by combining observations from air quality monitoring stations with both dispersion model output of the Long Term Ozone Simulation–EURopean Operational Smog model (LOTOS-EUROS) [38] and satellite data of aerosol optical thickness acquired by the Moderate Resolution Imaging Spectroradiometer (MODIS). Similarly, van de Kasteele [25] used kriging with external drift to combine NO_x ($NO_x = NO + NO_2$) observations with the output from a dispersion model. They found that a reduction in the density of the reference monitoring network for air quality can be compensated by geostatistical methods. Tilloy et al. [46] used a Kalman filter approach to assimilate air quality observations from 9 fixed stations into an urban air quality model. Applying a model covariance function that heavily relies on the shape of the street network, they found that the root-mean-square error of the assimilated fields is reduced by 30–50% over the model fields when station density is high. Silibello et al. [41] compared two data assimilation approaches, namely the successive corrections methods and optimal interpolation, to combine air quality observations with the output of a regional air quality model. Johansson et al. [20] described an infrastructure for performing data fusion of meteorological and air quality data for local and regional domains and found that hourly local NO_2 concentrations can be estimated more accurately with data fusion techniques than with conventional extrapolation methods.

Denby et al. [8] assessed the performance of two data assimilation methods used together with the LOTOS-EUROS model [38] for assessing PM_{10} exceedances on the European scale. The compared methods included the EnKF [10] and spatial interpolation based on a combination of residual kriging and linear regression. They found that assimilating the observations based on the spatial interpolation technique provided significantly superior results over those obtained from EnKF for their specific case study, with the root-mean-square error (RMSE) of the daily mean concentrations of PM_{10} at pre-selected validation stations being 9.2 $\mu g/m^3$ for the spatial interpolation technique and 13.5 $\mu g/m^3$ for EnKF. The poor performance of EnKF in this is case was primarily due to model bias which the used bias correction scheme was not able to properly account for. The residual kriging technique was found to be much less sensitive to model bias. Based on these results, we here use the closely related universal kriging technique as our data fusion method. To our knowledge, no previous studies have applied data assimilation and data fusion techniques for combining data from an urban-scale dispersion model with a network of crowdsourced observations of air quality.

3 Performance of Low-Cost Microsensors for Air Quality Monitoring

While there has been rapid and substantial progress in recent years with respect to the quality of low-cost microsensors for air quality applications, the quality of the data provided by the sensors is by no means close to that of traditional air quality monitoring stations equipped with reference instrumentation [2, 31, 6]. The data quality that can be expected from low-cost microsensors is often subject to significant systematic and random errors and thus to relatively low accuracy. While the systematic errors ("biases") can be to some extent corrected by co-locating the devices together with reference instrumentation of high accuracy, deriving the corresponding calibration equations and applying those to the original information, random errors cannot be corrected in such a way. The random error can be reduced, however, by performing temporal averaging or by averaging over multiple identical instruments. It should be noted that at this point there is no universally accepted test protocol for low-cost air quality sensors available; however, efforts along these lines are in progress within the European Commission through CEN (Comité Européen de Normalisation). In 2015, CEN/TC 264 WG42 (i.e., working group 42) started drafting documents describing specific performance requirements and ad hoc test methods for low-cost sensors and sensor arrays used for the determination of concentrations of gaseous pollutants and particulate matter in ambient air.

First studies evaluating the potential of the rapid technological advances in sensor technology for applications in air quality monitoring are beginning to be published but still remain somewhat limited in number [2, 16, 18, 19, 22, 27, 32, 33, 44, 45]. We have evaluated the performance of a set of commercially available low-cost air quality sensor platforms [6]. For this purpose, we used AQMesh units (Environmental Instruments Ltd, UK, www.aqmesh.com), which are battery operated and measure several gases including nitrogen monoxide (NO), nitrogen dioxide (NO_2), ozone (O_3), and carbon monoxide (CO) and the total particle count. In addition to laboratory evaluation, we have conducted a field test by co-locating a set of 24 identical AQMesh units at the Kirkeveien air quality monitoring station in the city of Oslo, Norway, between April and June 2015. The Kirkeveien air quality monitoring station is equipped with consistently maintained CEN approved gas analyzers, which offer highly reliable reference measurements to compare the observations from the low-cost sensors to.

While the evaluation results varied significantly from unit to unit, the strongest correlation among all measured species was found for NO, which exhibited Pearson correlation coefficients between 0.60 and 0.98 with an average of 0.86. Due to sensor interferences with ozone, NO_2 showed lower correlations of between 0.21 and 0.72 with an average of around 0.49.

One of the major issues of low-cost microsensors at this point is that they can have strong dependencies on the meteorological conditions, and particularly on temperature and relative humidity. Figure 2 shows the absolute bias, i.e., the difference between the sensor measurement of NO and the corresponding observations at the air quality monitoring station for a set of 24 AQMesh pods as a function of air temperature. In an ideal case, each panel would show bias values of zero across the entire temperature range. However, in practice, the behavior of the sensors varies significantly. While some nodes, such as 688150 or 744150, are pretty close to the ideal case, many of the other nodes show an increasing bias with temperature.

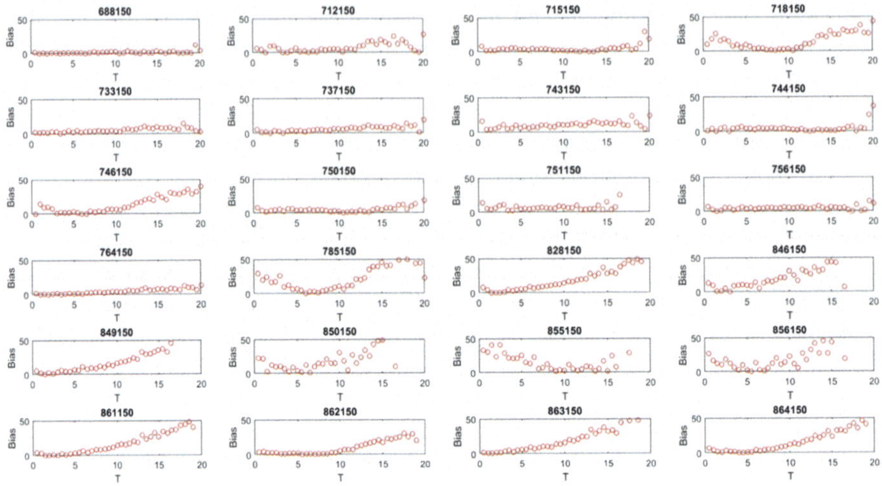

Fig. 2 Absolute bias of NO concentration (in units of µg/m^3) as a function of air temperature. All sensor nodes were identical and co-located at the same air quality monitoring station during the same period (April to June 2015). From [6]

Others, as for example 855150, even exhibit the opposite behavior with high biases at low temperatures and decreased biases with increasing temperatures.

While a temperature dependency in general can be corrected relatively easily, and in fact many manufacturers already carry out such a correction before shipping their products, in practice our real-world evaluation indicates that remaining temperature dependency varies considerably from sensor to sensor. It is therefore not possible to establish a field calibration equation for an individual node and then apply this to the rest of the nodes. Instead, each node has to be calibrated in the field to ensure optimal performance.

The air quality directive 2008/50/EC [9] defines the legal framework for air quality standards that all EU countries have to comply with. While primarily focused on reference air quality monitoring stations, it also describes supplementary techniques such as modeling and indicative measurements. In order to be considered as an indicative measurement according to the air quality directive, measurements need to comply with the Data Quality Objective (DQO), which provides a measure of acceptable uncertainty for indicative measurements (currently 25% for NO_2, NO_x, SO_2, and CO). Based on the methodology developed by Spinelle et al. [44], Fig. 3 shows an example of how the usability for indicative purposes can vary drastically between sensors. It provides the relative expanded uncertainty for NO of two identical sensor pods in comparison with the DQO for indicative measurements. While sensor pod 688150 reaches the DQO at concentrations of ca. 55 ppb, showing that it can be used for indicative measurements above this threshold, sensor pod 751150 never actually reaches the DQO threshold but remains at

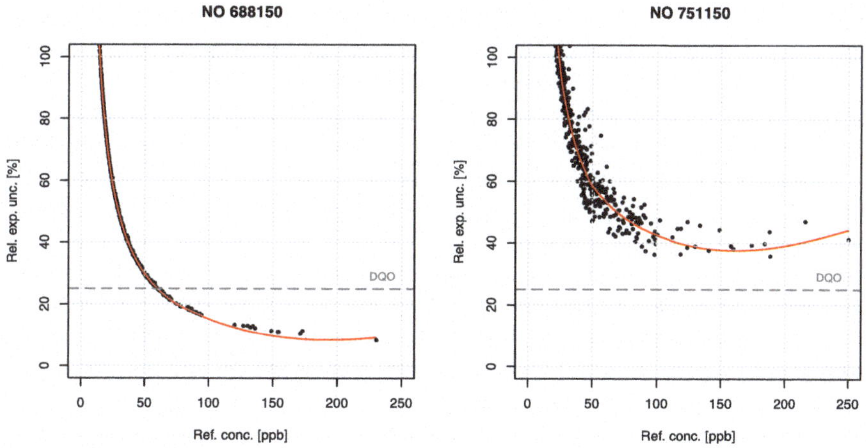

Fig. 3 Relative expanded uncertainty of NO in percent as a function of concentrations measured at a station equipped with reference instrumentation (Kirkeveien air quality monitoring station located in Oslo, Norway) for two identical low-cost air quality sensors. The dashed gray line represents the Data Quality Objective for indicative measurements of 25% as defined by the EU air quality directive [9]. The red line is a Loess fit to the data. Figure from [6]

relatively high uncertainties of around 40% even for quite high reference concentrations.

Data quality of low-cost microsensors is a pertinent concern, especially for citizen science application. We found that in a large set of identical sensors, the sensor performance varies significantly both spatially and temporally, is highly dependent on the environmental conditions, and finally varies drastically from sensor to sensor, making it necessary to carefully evaluate the data quality individually for each sensor pod. Nonetheless, it is possible to extract useful signals about air pollution from an entire network of low-cost sensors, as we will show in the next section. More information about the testing methodology and more detailed results of the evaluation can be found in Castell et al. [6].

4 Mapping Urban Air Quality Using Low-Cost Microsensors

Once (a) the expected accuracy of a set of low-cost air quality sensors has been established, (b) a thorough field calibration has been carried out, and (c) the sensors have been deployed at suitable locations throughout the domain of interest, the data provided by the sensor network can be exploited for applications in air quality monitoring and mapping. One of the most promising applications of such a sensor network is the production of up-to-date air quality maps of the urban environment at high spatial resolution (e.g., 100 m or less) and high frequency (e.g., hourly or less).

Similar information has been available in the past only from urban-scale dispersion models which tend to frequently suffer from significant biases as they are dependent on often highly uncertain local emission estimates.

We have exploited a network consisting of a total 24 AQMesh pods of low-cost microsensors located at the premises of kindergartens throughout the city of Oslo, in order to produce hourly air quality maps at relatively high spatial resolution (100 m). Despite the comparatively high spatial deployment density of a network of low-cost microsensors, it is not sufficient to simply interpolate the observations in space as the spatial gradients of the underlying concentration field are generally quite steep and would not be captured adequately by simple spatial interpolation. Such steep spatial gradients stem from the fact that traffic-related emissions are one of the major sources of air pollution in the urban environment and thus the highest concentrations tend to be found along the major road networks. In order to overcome this issue, we combine the air pollution observations from the network of low-cost microsensors with a constant or dynamic spatially exhaustive concentration field provided by an air quality model. This model information, which represents the typical air pollution conditions in the domain, is then used as a priori information and is subsequently modified globally and locally according to the sensor observations. We use here the output from the air pollution dispersion model EPISODE [42] in order to create concentration fields of the annual average concentration of the pollutant of interest (see Fig. 4). However, the method has also been successfully used with long-term average maps derived from land-use regression techniques, e.g., [15].

The time-invariant model-derived concentration field is then used an as a proxy dataset to guide the interpolation of the sensor observations using geostatistical techniques [7, 11, 17]. More specifically, we use universal kriging to carry out this task. This method is able to add value to both input datasets, namely the observations from low-cost sensors and the model information, by interpolating the point-based observations in a mathematically objective way while at the same time correcting the model information with true observations. The resulting concentration field is the best linear unbiased estimate (BLUE) of the true concentration field.

Mathematically, the combination of the two datasets is carried out by computing the estimated concentration $\hat{Y}(s_0)$ at point s_0 as

$$\hat{Y}(s_0) = c + a_1 \cdot x_1(s_0) + a_2 \cdot x_2(s_0) + \ldots + a_p \cdot x_p(s_0) + \varepsilon(s_0)$$

where c is a constant, a_1, a_2, etc. are regression coefficients, x, x_2, $\ldots x_p$ are the values of the p predictor variables of the regression component, and ε is a stationary random process with a given semivariogram. In matrix notation, we have

$$Y = \begin{bmatrix} Y_1 \\ \vdots \\ Y_n \end{bmatrix} = \begin{bmatrix} 1 & x_1(s_0) & \cdots & x_p(s_0) \\ 1 & \vdots & \ddots & \vdots \\ 1 & x_1(s_n) & \cdots & x_p(s_n) \end{bmatrix} \begin{bmatrix} c \\ a_1 \\ \vdots \\ a_p \end{bmatrix} + \begin{bmatrix} \varepsilon_1 \\ \vdots \\ \varepsilon_n \end{bmatrix} = Xa + \varepsilon$$

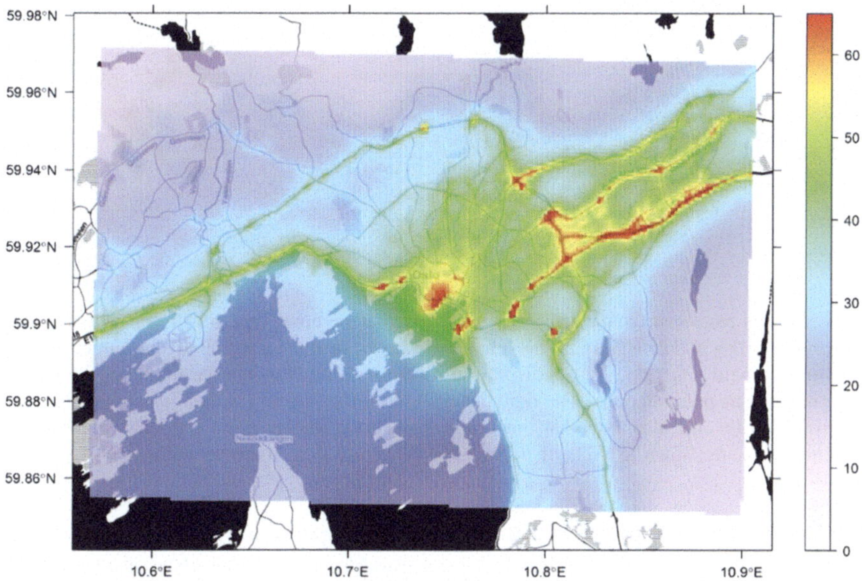

Fig. 4 Annual average NO_2 concentration of 2014 for Oslo, Norway, as derived from the output of the EPISODE air pollution dispersion model [42]. Concentrations are shown in units of $\mu g/m^3$

where Y indicates the estimated values at all prediction locations, X represents the values of the predictor variables at all locations, a is the vector of regression coefficients, ε indicates the vector of residual errors that is estimated using kriging with the known semivariogram model, n is the number of prediction locations, and p is the number of predictor variables. In practice, only a single predictor variable (the modeled spatially exhaustive concentration field) is generally used. More details about the specific methodology can be found in previous studies [30, 39, 40].

Using the method described above, we performed data fusion of the network of 24 AQMesh sensors deployed throughout Oslo with a time-invariant annual average concentration map provided by the EPISODE model. In order to illustrate the basic principle of the data fusion methodology, Fig. 5 shows the two input datasets to the method (left panel) and the output of the data fusion (right panel) after combining the two input datasets for a single hour (February 1, 2016 at 12:00 UTC). When comparing the two input datasets, it can be seen that the colors between the background map and the point markers tend to differ quite substantially. This is because the selected day is a relatively highly polluted winter day, so the observed concentrations acquired by the sensors tend to overall be higher than the annual average concentrations predicted by the model. Once these two datasets are fused using the methodology described above (background map of the right panel with the point markers shown for reference), it can be seen that the fusion

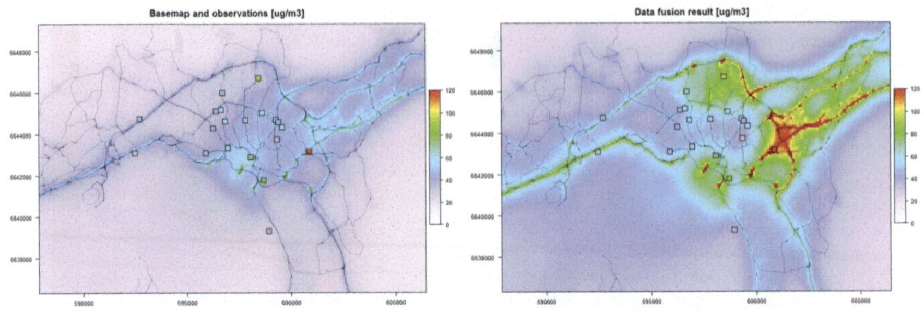

Fig. 5 Typical results of the data fusion process for a single hour, here shown for NO_2 on the 1st of February 2016 at 12:00 UTC. The left panel shows in the background the annual average concentration field generated by the dispersion model and the NO_2 observations acquired by the sensor network as point markers. The right panel shows the results of the data fusion algorithm as a background map and also includes the original observations as point markers for reference. Concentrations are shown in units of $\mu g/m^3$. The thin gray lines indicate major roads

technique manages to preserve the overall typical spatial air pollution patterns and the spatial gradients of the model-derived annual average concentration map. At the same time, the method adjusts the a priori concentration field derived from the model locally in the immediate surroundings of the sensors in such a way that the concentration field matches the observations. A good example is the red hotspot measured on the eastern side of the city. The concentration measured here significantly exceeded the annual average prediction from the model and as such the data fusion map if significantly corrected upwards in this area. It should be noted that, since this observation was made relatively close to a road, the resulting correction increases primarily the concentration along the road network and only to a lesser extent the background concentrations further away from the roads.

Figure 6 shows an example of a full day of hourly data fusion output for NO_2. We can observe how the overall spatial patterns of typical air pollution are preserved in the various maps; however, they are adjusted both globally and locally based on the observations from the low-cost microsensors. The daily cycle of NO_2 pollution, which is primarily caused by traffic emissions, is clearly visible in Fig. 6. Particularly the morning rush hour is quite evident as we can observe a sharp increase in concentrations at around 7:00 local time (6:00 UTC), which then continues to rise until concentrations reach the daily maximum around 9:00 local time (8:00 UTC). After the distinct morning rush hour, overall concentration decreases slightly and then remains more or less stable throughout the day. Evening rush hour is less visible as it usually tends to spread over a longer time range; however, slightly elevating concentrations are again visible between 17:00 and 19:00 local time (16:00 to 18:00 UTC). After 21:00 local time (20:00 UTC), city-wide concentration of NO_2 drops significantly and reaches the daily minimum around midnight and remains at these low levels throughout the night.

It should be noted here again that the dispersion model is only used for creating a constant annual average map of NO_2 (or other pollutants), which indicates the

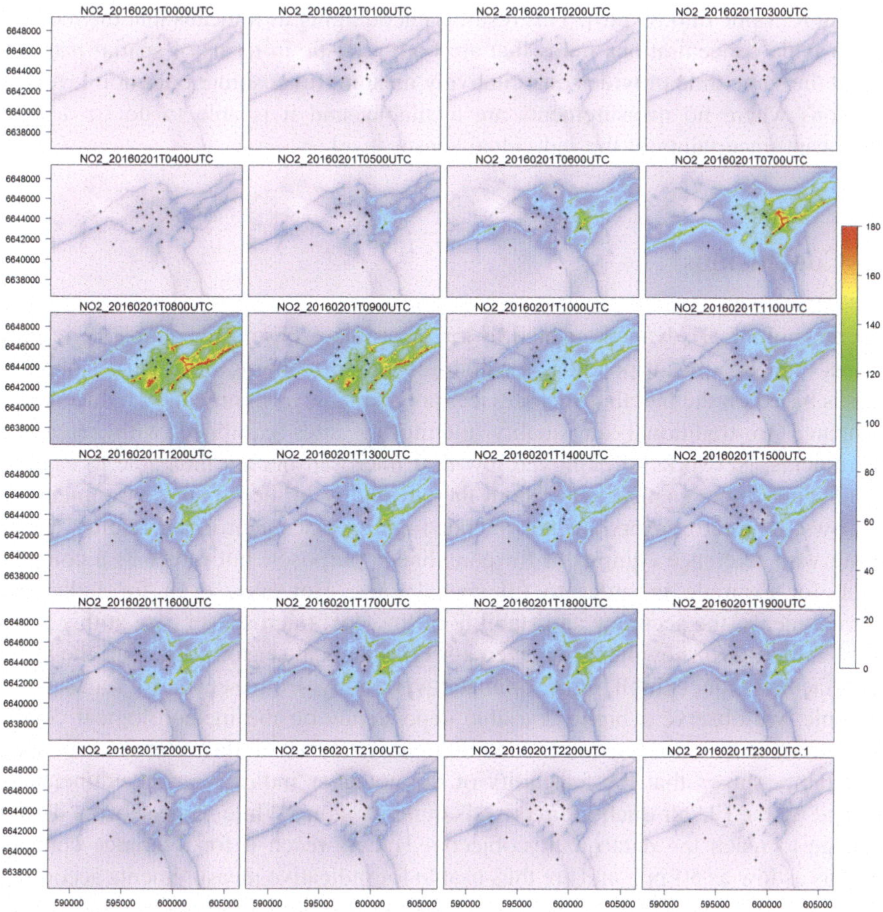

Fig. 6 Example of a full day of hourly data fusion results for NO_2, here shown for 1st of February 2016. Concentrations are shown in units of µg/m^3. *Note* that the time is given here in UTC, whereas local time is CET (CET = UTC + 1 h). The axes show the map coordinates in UTM projection (Zone 32, WGS84). Black crosses indicate the locations of the deployed AQMesh sensors pods

spatial patterns of the typical pollution levels throughout the city. These spatial patterns are generally inherited by the fused map. However, the daily cycle of pollution that can be seen in Fig. 6 is entirely based on the observed concentrations from the network of low-cost microsensors. Quantitatively, when the fused predictions are extracted from the maps at the locations of the official air quality monitoring stations, we found as part of a validation exercise that the maps resulting from data fusion are capable of reproducing the average measured NO_2 concentrations at the stations (where no low-cost microsensors were deployed) with a mean bias of 5 µg/m^3, an RMSE value of 14.3 µg/m^3 and a quite robust correlation with an R^2 value of 0.89 [40]. This result is encouraging as it means that the method can provide concentrations fields that are both realistic in terms of spatial patterns and at the same time provides quantitatively meaningful absolute pollution levels at locations where no measurements are available, and it is able to do so despite significant uncertainty at the individual sensor level.

5 Conclusions

The technological advances related to low-cost microsensors for atmospheric pollutants offer significant potential for high-resolution mapping of urban air quality. This is based on the fact that such devices are available at significantly reduced cost compared to traditional air quality monitoring sites equipped with reference instruments. Networks of deployed low-cost microsensors are thus able to capture significantly more spatial detail about the air quality in the urban environment.

Low-cost microsensors are not intended as a replacement of air quality monitoring with reference equipment for compliance purposes, but rather as a complement, for example in order to fill spatial gaps. However, even for indicative measurements, the accuracy and stability of low-cost microsensors are at this point quite limited. We show here that even a set of identical sensor pods from the same manufacturer can exhibit quite dramatic variability from sensor to sensor. For example, we observe a highly variable dependency on the meteorological conditions from sensor to sensor. In addition, an analysis of the relative expanded uncertainty shows that the feasibility of the nodes as indicative measurements as defined by the EU air quality directive is quite variable. While some sensors do not manage to reach the data quality objective, others reach it for reference concentrations as low as 50 ppb and are thus usable for indicative measurements according to guidelines of the EU directive, at least when the pollution levels are moderate to high. At very low pollution levels, the signal-to-noise ratio of the majority of sensors currently prohibits their use for critical regulatory applications.

Nonetheless, despite the significant challenges with individual sensor units, under the right conditions and with proper calibration as well as post-processing, it is already possible to exploit the data delivered by networks of low-cost sensors for mapping urban air quality. We show how a relatively simple method based on geostatistics is able to combine the information from the sensor network with a time-invariant map of air quality and is thus able to provide up-to-date maps of urban air quality that objectively interpolate the sensor observations while at the same time constraining the model information with true measurements. The crucial point to keep in mind here is that the method makes use of the "swarm knowledge" of the entire sensor network rather than relying on the data from an individual sensor, which are often unreliable by themselves. As the presented method implicitly tends to exploit an entire cluster of neighboring sensor observations to map the overall pollution levels for a neighborhood, individual sensors have limited weight on the resulting concentration field. As such, it is rather an ensemble of sensors that determines the local modification of the annual average concentration field. The resulting data fusion maps reflect this as well as the fact that areas without any observations contain spatial detail inherited from the model.

The rapid current pace of technological development is likely to continue in the near future and significant advances in sensor technology are expected in the upcoming years, resulting in low-cost air quality sensors with lower systematic and random errors and as such dramatically improved overall accuracy as well as improved sensor stability and reduced sensor-to-sensor variability. The latter two points in particular will be crucial for a further exploitation of networks of air quality sensors as they will allow for simplified calibration procedures compared to the status quo. In addition, the production costs of the sensors are likely to see further substantial reductions in the near future, thus enabling even larger-scale deployments of air quality sensor networks than currently possible, which in turn will result in further increased spatial sampling densities throughout the urban environment.

All of these developments in combination will likely lead to a global increase in the uptake of low-cost microsensor networks for air quality monitoring, primarily as a complementary technique to traditional air quality monitoring stations equipped with reference instrumentation. Furthermore, while the use of such sensor networks for high-resolution mapping of urban air quality under ideal and controlled conditions was already demonstrated here, the described advances will significantly increase the usefulness of data fusion and data assimilation techniques for high-resolution mapping of urban air quality in the future.

Acknowledgements The work described here was carried out with funding provided by the European Union's Seventh Framework Programme for research, technological development and demonstration under grant agreement no. 308,524 (CITI-SENSE). Additional support was provided by the Citi-Sense-MOB project partially funded by EMMIA: The European Mobile and Mobility Industries Alliance under grant agreement no. SI2.647655.

References

1. Aleixandre, M., & M. Gerboles (2012). Review of small commercial sensors for indicative monitoring of ambient gas review of small commercial sensors for indicative monitoring of ambient gas. Chemical Engineering Transcation *30*, doi:https://doi.org/10.3303/CET1230029.
2. Borrego, C., et al. (2016). Assessment of air quality microsensors versus reference methods: The EuNetAir joint exercise. *Atmospheric Environment, 147*(2), 246–263. https://doi.org/10.1016/j.atmosenv.2016.09.050.
3. Van den Bossche, J., Peters, J., Verwaeren, J., Botteldooren, D., Theunis, J., & De Baets, B. (2015). Mapping spatial variation in urban air quality using mobile measurements: development and validation of a methodology based on an extensive dataset. *Atmospheric Environment, 105,* 148–161. https://doi.org/10.1016/j.atmosenv.2015.01.017.
4. Bouttier, F., & Courtier P. (1999, March). Data assimiloation concepts and methods.
5. Castell, N., Kobernus, M., Liu, H.-Y., Schneider, P., Lahoz, W., Berre, A. J., et al. (2014). Mobile technologies and services for environmental monitoring: The Citi-Sense-MOB approach. *Urban Clim., 14,* 370–382. https://doi.org/10.1016/j.uclim.2014.08.002.

6. Castell, N., Dauge, F. R., Schneider, P., Vogt, M., Lerner, U., Fishbain, B., et al. (2017). Can commercial low-cost sensor platforms contribute to air quality monitoring and exposure estimates? *Environment International, 99,* 293–302. https://doi.org/10.1016/j.envint.2016.12.007.
7. Chilès, J.-P., & Delfiner P. (2012). Geostatistics: Modeling spatial uncertainty. John Wiley & Sons.
8. Denby, B., Schaap, M., Segers, A., Builtjes, P., & Horálek, J. (2008). Comparison of two data assimilation methods for assessing PM10 exceedances on the European scale. *Atmospheric Environment, 42*(30), 7122–7134. https://doi.org/10.1016/j.atmosenv.2008.05.058.
9. EU. (2008). Directive 2008/50/EC of the European Parliament and the Council of 21 May 2008 on Ambient Air Quality and Cleaner Air for Europe.
10. Evensen, G. (2003). The Ensemble Kalman filter: Theoretical formulation and practical implementation. *Ocean Dynamics, 53*(4), 343–367. https://doi.org/10.1007/s10236-003-0036-9.
11. Goovaerts, P. (1997). *Geostatistics for natural resources evaluation*. New York: Oxford University Press.
12. Guerreiro, C., Gonzalez Ortiz, A., de Leeuw, F., Viana, M., & Horalek, J. (2016). *Air quality in Europe—2016 report*. Copenhagen, Denmark: European Environment Agency.
13. Hasenfratz, D., O. Saukh, S. Sturzenegger, & L. Thiele (2012). Participatory air pollution monitoring using smartphones. In: Proceedings of the 2nd international workshop on mobile sensing, April 16–20, 2012, Beijing, China, pp. 1–5.
14. Hasenfratz, D., O. Saukh, C. Walser, C. Hueglin, M. Fierz, T. Arn, J.et al. (2014). Deriving high-resolution urban air pollution maps using mobile sensor nodes. *Pervasive and Mobile Computing, 16*(Part B), 268–285. doi:https://doi.org/10.1016/j.pmcj.2014.11.008.
15. Hoek, G., Beelen, R., de Hoogh, K., Vienneau, D., Gulliver, J., Fischer, P., et al. (2008). A review of land-use regression models to assess spatial variation of outdoor air pollution. *Atmospheric Environment, 42*(33), 7561–7578. https://doi.org/10.1016/j.atmosenv.2008.05.057.
16. Holstius, D. M., Pillarisetti, A., Smith, K. R., & Seto, E. (2014). Field calibrations of a low-cost aerosol sensor at a regulatory monitoring site in California. *Atmospheric Measurement Techniques, 7*(4), 1121–1131. https://doi.org/10.5194/amt-7-1121-2014.
17. Isaaks, E. H., & Srivastava, R. M. (1989). *Applied geostatistics*. New York: Oxford University Press.
18. Järvinen, A., Kuuluvainen, H., Niemi, J. V., Saari, S., Dal Maso, M., Pirjola, L., et al. (2015). Monitoring urban air quality with a diffusion charger based electrical particle sensor. *Urban Clim., 14,* 441–456. https://doi.org/10.1016/j.uclim.2014.10.002.
19. Jiao, W., et al. (2016). Community Air Sensor Network (CAIRSENSE) project: Evaluation of low-cost sensor performance in a suburban environment in the southeastern United States. *Atmospheric Measurement Techniques, 9*(11), 5281–5292. https://doi.org/10.5194/amt-9-5281-2016.
20. Johansson, L., Epitropou, V., Karatzas, K., Karppinen, A., Wanner, L., Vrochidis, S., et al. (2015). Environmental Modelling & Software Fusion of meteorological and air quality data extracted from the web for personalized environmental information services. *Environmental Modelling and Software, 64,* 143–155. https://doi.org/10.1016/j.envsoft.2014.11.021.
21. Journel, A. G., & Huijbregts, C. J. (2003). Mining geostatistics. Blackburn Press.
22. Jovašević-Stojanović, M., Bartonova, A., Topalović, D., Lazović, I., Pokrić, B., & Ristovski, Z. (2015). On the use of small and cheaper sensors and devices for indicative citizen-based monitoring of respirable particulate matter. *Environmental Pollution, 206,* 696–704. https://doi.org/10.1016/j.envpol.2015.08.035.
23. Kalman, R. E. (1960). A new approach to linear filtering and prediction problems. *Journal of basic Engineering, 82*(Series D), 35–45.
24. Kalnay, E. (2003). *Atmospheric modeling data assimilation and predictability*. Cambridge, UK: Cambridge University Press.

25. van de Kassteele, J., Stein, A., Dekkers, A. L. M., & Velders, G. J. M. (2009). External drift kriging of NOx concentrations with dispersion model output in a reduced air quality monitoring network. *Environmental and Ecological Statistics, 16*(3), 321–339. https://doi.org/10.1007/s10651-007-0052-x.
26. Van De Kassteele, J., Koelemeijer, R. B. A., Dekkers, A. L. M., Schaap, M., Homan, C. D., & A. Stein (2006), Statistical mapping of PM10 concentrations over Western Europe using secondary information from dispersion modeling and MODIS satellite observations, *Stochastic Environmental Research and Risk Assessment, 21*(2), 183–194. doi:https://doi.org/10.1007/s00477-006-0055-4.
27. Kelly, K. E., Whitaker, J., Widmer, C., Dybwad, A., & Butterfield, A. (2017). Ambient and laboratory evaluation of a low-cost particulate matter sensor (submitted). *Environmental Pollution, 221,* 491–500. https://doi.org/10.1016/j.envpol.2016.12.039.
28. Kumar, P., Morawska, L., Martani, C., Biskos, G., Neophytou, M., Di Sabatino, S., et al. (2015). The rise of low-cost sensing for managing air pollution in cities. *Environment International, 75,* 199–205. https://doi.org/10.1016/j.envint.2014.11.019.
29. Lahoz, W., Khattatov, B., & Menard, R. (Eds.). (2010). *Data Assimilation*. Berlin, Heidelberg: Springer.
30. Lahoz, W. A., & Schneider, P. (2014). Data assimilation: Making sense of Earth Observation. *Frontiers in Environmental Science, 2*(16), 1–28. https://doi.org/10.3389/fenvs.2014.00016.
31. Lewis, A., & Edwards, P. (2016). Validate personal air-pollution sensors. *Nature, 535,* 29–31.
32. Mead, M. I., et al. (2013). The use of electrochemical sensors for monitoring urban air quality in low-cost, high-density networks. *Atmospheric Environment, 70,* 186–203. https://doi.org/10.1016/j.atmosenv.2012.11.060.
33. Moltchanov, S., Levy, I., Etzion, Y., Lerner, U., Broday, D. M., & Fishbain, B. (2015). On the feasibility of measuring urban air pollution by wireless distributed sensor networks. *Science of the Total Environment, 502,* 537–547. https://doi.org/10.1016/j.scitotenv.2014.09.059.
34. Nieuwenhuijsen, M. J., Donaire-Gonzalez, D., Rivas, I., De Castro, M., Cirach, M., Hoek, G., et al. (2015). Variability in and agreement between modeled and personal continuously measured black carbon levels using novel smartphone and sensor technologies. *Environmental Science and Technology, 49*(5), 2977–2982. https://doi.org/10.1021/es505362x.
35. Peters, J., Theunis, J., van Poppel, M., & Berghmans, P. (2013). Monitoring PM10 and ultrafine particles in urban environments using mobile measurements. *Aerosol and Air Quality Research, 13,* 509–522. https://doi.org/10.4209/aaqr.2012.06.0152.
36. Peters, J., Van den Bossche, J., Reggente, M., Van Poppel, M., De Baets, B., & Theunis, J. (2014). Cyclist exposure to UFP and BC on urban routes in Antwerp, Belgium. *Atmospheric Environment, 92,* 31–43. https://doi.org/10.1016/j.atmosenv.2014.03.039.
37. Piedrahita, R., Xiang, Y., Masson, N., Ortega, J., Collier, A., Jiang, Y., et al. (2014). The next generation of low-cost personal air quality sensors for quantitative exposure monitoring. *Atmospheric Measurement Techniques, 3*(2), 3325–3336. https://doi.org/10.5194/amt-7-3325-2014.
38. Schaap, M., Timmermans, R. M., Roemer, M., Boersen, G. A. C., Builtjes, P., Sauter, F. J., et al. (2008). The LOTOS? EUROS model: description, validation and latest developments. *International Journal of Environment and Pollution, 32*(2), 270–290. https://doi.org/10.1504/IJEP.2008.017106.
39. Schneider, P., Castell, N., Vallejo, I., Vogt, M., Lahoz, W., & Bartonova A.(2016). Data fusion of crowdsourced observations and model data for high-resolution mapping of urban air quality. In *10th International Conference on Air Quality—Science and Applications*, Milan, Italy.
40. Schneider, P., Castell, N., Vogt, M., Dauge, F. R., Lahoz, W. A., & Bartonova, A. (2017). Mapping urban air quality in near real-time using observations from low-cost sensors and model information. *Environment International, 106*(May), 234–247. https://doi.org/10.1016/j.envint.2017.05.005.

41. Silibello, C., Bolignano, A., Sozzi, R., & Gariazzo C. (2014). Application of a chemical transport model and optimized data assimilation methods to improve air quality assessment. *Air Quality, Atmosphere & Health*. doi:https://doi.org/10.1007/s11869-014-0235-1, doi: https://doi.org/10.1007/s11869-014-0235-1.
42. Slørdal, L. H., Walker, S.-E., & Solberg, S. (2003). *The urban air dispersion model EPISODE applied in AirQUIS 2003—technical description*. Norway: Kjeller.
43. Snyder, E. G., Watkins, T. H., Solomon, P. A., Thoma, E. D., Williams, R. W., Hagler, G. S. W., et al. (2013). The changing paradigm of air pollution monitoring. *Environmental Science and Technology, 47*(20), 11369–11377. https://doi.org/10.1021/es4022602.
44. Spinelle, L., Gerboles, M., Villani, M. G., Aleixandre, M., & Bonavitacola, F. (2015). Field calibration of a cluster of low-cost available sensors for air quality monitoring Part A: Ozone and nitrogen dioxide. *Sensors and Actuators B: Chemical, 215*, 249–257. https://doi.org/10.1016/j.snb.2015.03.031.
45. Spinelle, L., Gerboles, M., Villani, M. G., Aleixandre, M., & Bonavitacola, F. (2017). Field calibration of a cluster of low-cost commercially available sensors for air quality monitoring. Part B: NO, CO and CO_2. *Sensors and Actuators B: Chemical, 238*, 706–715. https://doi.org/10.1016/j.snb.2016.07.036.
46. Tilloy, A., Mallet, V., Poulet, D., Pesin, C., & Brocheton, F. (2013). BLUE-based NO2 data assimilation at urban scale. *Journal of Geophysical Research: Atmospheres, 118*(4), 2031–2040. https://doi.org/10.1002/jgrd.50233.
47. Wackernagel, H. (2003). *Multivariate Geostatistics*. Berlin Heidelberg: Springer.
48. World Health Organization. (2016). Ambient air pollution: A global assessment of exposure and burden of disease. World Health Organization.

The Urban Nexus Project: When Urban Mobility Analysis, VGI and Data Science Meet Together

How an Urban Mobility Analysis Project Fosters the Development of Data Science Technology

Federica Burini, Daniele E. Ciriello, Alessandra Ghisalberti and Giuseppe Psaila

Abstract This chapter discusses a multidisciplinary research project called *Urban Nexus*, which aims at studying the role of big data in urban governance, along with the need of a high-level query language to allow analysts, geographers and, in general, non-programmers to easily cross-analyze multisource *big data* produced by the inhabitants and coming also from other sources of information. The project focuses on the production of Volunteered Geographic Information by means of apps, crowdsourced data from social networks and fosters an analysis based also on authoritative geo-referenced data coming from multiple institutional sources. The *Urban Nexus* project applies novel and multidisciplinary approaches to interpret social aspects of big data, in relation to three medium-sized cities and their surrounding areas: *Bergamo* in the wider context of Milan (*Italy*), *Lausanne* and the other cities of *Switzerland* and *Cambridge* and the metropolitan area of London (*UK*). Focused on urban regeneration and mobility, we want to devise and test methods to make such cities *learning cities*, i.e., territories where the inhabitants help to identify the design guidelines according to their competence on places that is their *spatial capital* [1]. In particular, we identified an analysis method of data sets represented by means of the *JSON* lightweight data-interchange format.

F. Burini
Department of Foreign Languages, Literatures and Cultures, CST-DiathesisLab, University of Bergamo, Via Salvecchio 19, 24129 Bergamo, BG, Italy

D. E. Ciriello
CST-DiathesisLab, University of Bergamo, Via Salvecchio 19, 24129 Bergamo, BG, Italy

A. Ghisalberti
Department of Foreign Languages, Literatures and Cultures, University of Bergamo, Via Salvecchio 19, 24129 Bergamo, BG, Italy

G. Psaila (✉)
Department of Management, Information and Production Engineering, University of Bergamo, Viale Marconi 5, 24044 Dalmine, BG, Italy
e-mail: giuseppe.psaila@unibg.it

G. Bordogna and P. Carrara (eds.), *Mobile Information Systems Leveraging Volunteered Geographic Information for Earth Observation*, Earth Systems Data and Models 4, https://doi.org/10.1007/978-3-319-70878-2_6

Currently, *JSON* has become the de facto standard for exchanging Volunteered Geographic Information (VGI) and crowdsourced data created within social networks. Since an easy to use high-level language for querying and manipulating collections of possibly geo-tagged *JSON* objects is still unavailable, as well as an integrated framework for data management and visualization of geo-tagged JSON data sets, we devised a new framework named *J-CO*. In this framework, we propose a high-level language, called *J-CO-QL* (firstly introduced in [2]), which is based on a well-defined execution model. A plug-in for *QGIS* is provided as well: it permits to visualize geo-tagged data sets stored in a *NoSQL* database such as *MongoDB*; furthermore, the same plug-in can be used to write and execute *J-CO-QL* queries on those databases. By means of a practical example, we show how the *J-CO* framework is able to allow geographers and analysts to work on complex data sets coming from *VGI* sources, in order to study the mobility of city users who voluntarily participated to the gathering process of data for the city of *Bergamo*.

1 Introduction

Traditional approaches to social analysis rely on the observation of social phenomena through questionnaires, surveys and interviews. However, it is a matter of fact that many other sources of information describing social phenomena are now available and could be exploited. Social networks have shown that people wish to give their opinion on topics they perceive to be important. This fact suggested the concept of *Volunteered Info*rmation, i.e., information given by users in a voluntary manner. When volunteer information concerns geographical aspects, we obtain the so-called *Volunteer Geographical Information* (*VGI* for short) [3].

VGI is usually collected by means of mobile apps that often trace the movements of users and provide them with an interface to actively annotate their traces and the places they visited, in relation to their activities.

However, in order to conduct research activities and to apply modern approaches to the study of urban mobility and beyond, *VGI* is not enough. In particular, an interdisciplinary approach is mandatory, because many variegate competences are necessary. An interesting example is the *Urban Nexus* project, leaded by the *University of Bergamo*.

Evoking the *EU* action program *Urban Nexus* between European cities aimed at promoting an integrated approach to urban sustainability, the project aims to create a scientific and educational cooperation between the *University of Bergamo*, the *Ecole Polytechnique Fédérale de Lausanne* and *Anglia Ruskin University* in *Cambridge*. With the *Urban Nexus* project, we want to apply novel and multidisciplinary approaches to interpret social aspects of big data, in relation to three medium-sized cities (*Bergamo*, *Lausanne* and *Cambridge*) and their surrounding areas. Being focused on urban regeneration and mobility, we want to devise and test methods to make such cities *learning cities*, i.e., territories where the inhabitants help to identify the design guidelines according to their competence of places and their spatial capital.

In particular, the methodological approach we adopted is based on the integration of *VGI* with authoritative information that is publicly available on *Open Data Portal*s. In fact, in recent years, public administrations have to publish data sets concerning the territories they administrate. Many and various aspects of city users' mobility are described by open data sets. More in general, *Big Dat*a are now considered a valuable source of information for performing social and territorial analysis. The *Urban Nexus* project aims at exploring this frontier.

It is clear that traditional approaches to social and territorial analysis are no longer adequate: analysts must integrate and cross-analyze multiple and heterogeneous (in terms of both format and content) data sets, available in digital formats. Consequently, the support provided by computer science has become crucial, in the sense of having to deal with many and many data sets (this is the concept of *Variety* in the *3-Vs* and *4-Vs* models of *Big Data*). From the point of view of a computer programmer, everything can be done (in principle); however, analysts are not computer programmers and they want to reduce their dependency on computer programmers, when they perform their analysis. On the opposite side, computer scientists do not have the knowledge related to spatial analysis and social phenomena, useful to address and guide the choice of data to be collected and the reasons to collect them.

In practice, computer science must provide high-level tools that could be used by non-programmer analysts (social scientists and geographers) to perform their geo-referenced analyses. Such tools can be classified as *data science* tools, because they should be useful for data transformation and analysis. Obviously, from a computer science perspective, tools developed for data science should not be specifically devised for a specific task or problem; in contrast, they should be devised in order to be suitable for the broadest application scenarios, although they are inspired by specific needs.

In the *Urban Nexus* project, the methodological approach is to involve city users in the gathering of data through *VGI*. This is obtained by means of the app named *Moves*, a powerful app available for *Android* and *iOS* mobile platforms, that traces movements of users, by recording their *GPS* traces; furthermore, users could also annotate their traces or places they visited. The aim of the project is to cross-analyze *VGI* with authoritative data sets concerning public transportations.

Since the early stage of the project, whose members are geographers, spatial analysts, and computer scientists, clearly emerged the need for a framework for managing the collected data sets. In particular, having noticed that the *JSON* lightweight data-interchange format has become the de facto standard for data coming from social networks and mobile app *API*s, we understand that a data management framework to store, transform, cross-analyze and visualize possibly geo-tagged *JSON* data sets is necessary.

The result of this work, from a data science and computer science perspective, is the design of the *J-CO Framework*. This framework provides the capability of managing collections of *JSON* data sets stored in a *NoSQL* database such as

MongoDB, a query language (named *J-CO-QL*) that permits to specify complex transformations at a high level, and a plug-in for *QGIS*, the free and well-known *GIS* software.

By means of a practical example, we show how the *J-CO* framework is able to allow geographers and analysts to work on complex data sets coming from real *VGI* sources, in order to study city users that voluntarily participated to the *VGI* gathering process for the city of *Bergamo*.

In our opinion, the *Urban Nexus* project is an example of how a multidisciplinary project can foster the development of data science tools for a large variety of data science applications.

This chapter is organized as follows: In Sect. 2, we present the project and the methodological approach, in particular as far as the involvement of city users and the *VGI* gathering are concerned. In Sect. 3, we present the *J-CO* Framework and its main characteristics. In Sect. 4, we present how *J-CO-QL* can be effectively used by non-programmers, in order to transform *VGI* data and detect city users reaching the city by train. Section 5 draws the conclusions.

2 Urban Nexus—Intelligent Modeling and Big Data Mapping for the Analysis of Connectivity and Regeneration of European Cities

Urban Nexus aims to create a scientific and educational cooperation between the *University of Bergamo*, the *Ecole Polytechnique Fédérale de Lausanne* and *Anglia Ruskin University* in *Cambridge* with the overall objective to analyze, through an integrated approach, measures intended to foster the material renovation and the improvement of accessibility of the three cities and surrounding areas involved, in a comparative way and in the framework of the "*contemporary city*" [4]. The project adopts an approach that interprets the "*city*" as a node of a globalized network where there are no more a local and a global dimension, but rather these dimensions interact with reconfiguring urban contexts with their centralities, their axis, their full and empty spaces, and their internal and external connections [5]. In fact, the regeneration of the built environment, intertwined with that of its accessibility, is a common problem for many European cities. The so-called "*city centers*" are public spaces that need revitalization, buildings are empty and shops and offices are in superabundance; those who once were called "*suburbs*" have the demand for new housing, rehabilitation services, as well as a response to soil consumption as a consequence of the presence of a large number of obsolete and abandoned buildings. For this reason, the "*places*" of living, of living in a city, cannot be divided into neighborhoods, but should be analyzed through the concept of *network*, as it better grasps its significance in relation to the inhabitant needs and movements. In short, if the base of contemporary life is movement, the element from which to explore the city is its people (stakeholders of mobility) and its places as nodes of a network that creates unity and cohesion [6].

From the methodological point of view, relying on the potential of *Big Data* [7] and by exploiting intelligent modeling techniques and big data mapping systems, we analyze the function and the use of urban spaces based on the needs of the inhabitants and of city users. To this end, the project examines the three urban locations of the universities involved (*Bergamo*, *Lausanne,* and *Cambridge*), cluster by virtue of demographic size, connectivity through high-speed transportation, the presence of a university system. We are analyzing data sets from national statistics and big data from private sources (social media or mobile phone companies), offering their analysis, and reaching their interpretation by *cyber-cartographic systems* aimed at understanding the habits, the flows and relationships of the inhabitants and of the city users (residents, commuters, tourists, migrants, and so on). These are released by their connotations of residency and related to different ways of experiencing the world and of managing the distances within it [8]. It is a research already introduced in *Europe* by the *European Statistical System* and accepted by the *Italian National Statistics*, particularly from the "*Commission to study and guide the choices of ISTAT on Big Data,*" which is involved in the project, together with *IREA-CNR* of *Milan*. Thus, the goal is go beyond classical cartography [9], in order to think about space in a new way.

2.1 Organization of the Activities

The *Urban Nexus* project aims at performing several research activities that involve researchers with different competences, mainly, geographers, spatial analysts and computer engineers. In fact, the large amount of data to collect and cross-analyze, demands for computer engineers able to provide flexible tools able to deal with a large number of possibly huge data sets, coming with different formats and heterogeneous structures. Nevertheless, the choice and analysis of data must be carried on and the results interpreted by geographers and spatial analysts.

The activities of the project can be summarized as follows:

- ***socio-spatial domain***: addressed to the identification of the peculiarities of urban spaces investigated in a reticular and polycentric perspective, the analysis of the socio-spatial dynamics in place in the three cities and the identification of the city users and their mobility, the creation of participatory and VGI processes aimed at consultation and discussion with the stakeholders of the investigated areas, the study of the potential regeneration of brownfields and obsolete areas in the three cities;
- ***intelligent modeling***: addressed to the various activities related to the collection, processing, synthesis, and analysis of *Big Data* by: data mining, machine learning, spatial-temporal data analysis;
- ***mapping***: addressed to the conception and design of models of representation of urban connectivity; the creation of systems of collaborative web mapping [10];

- ***computer graphics and teaching support systems***: development of interactive systems and media for communication of research results; development of information systems for teaching.

2.2 City Users Detection and Analysis

In the project, one key activity is to identify city users. For this reason, we decided to launch a *VGI* (*Volunteered Geographical Information*) and participatory process, in order to obtain the help of typical city users, i.e., students of high schools and of the *University of Bergamo*. By organizing focus groups held by the researchers and the students involved in the Urban Nexus project, we are asking them to download the app named *Moves* and to register to it; then, they are asked to register to our web site, by providing the registration code given by *Moves*, so as to be able to automatically download their traces. In fact, *Moves* is a mobile application branded by *FaceBook*, which registers movements of users, distinguishing the mode of transport: when they move on foot or by car, for example. Users can obtain data regarding their own movements, possibly viewing them on a map service. Nevertheless, *APIs* are provided to developers that can access traces of registered users.

Thus, a specific task is to identify city users that reach *Bergamo* by train. This task is useful to understand the technical context and the need for the *J-CO* framework.

To conduct the task of identifying people who usually travel by public transports, in particular, by train, we need to collect data from various sources, i.e., *Open Data* portals, official data about public transportations provided by transport companies, and *Moves*. We then cross-analyze those heterogeneous data sets in order to obtain simple results and visualize those results in our favorite *GIS* environment. Figure 1 depicts the way the task is carried-on: data sets are collected and stored into a storage service; then, they are cross-analyzed and the results of this cross-analysis are shown in a *GIS* software, for visual analysis of results.

Notice the heterogeneity of the data sets: data coming from *Moves* (*JSON* data set having a very complex structure); data describing the timetable of public transports and the lines (possibly *CSV* or *JSON* data sets); data about position of train stations (open data possibly in *CSV* or *JSON*).

The cross-analysis of these heterogeneous data sets could be performed (with certain limitations) by skilled programmers. However, this would require to write new programs for each analysis to perform, impeding geographers from actively participating to the analysis process. In contrast, geographers wish to be active while performing analysis: to do that, they need a data management framework that is able to deal with heterogeneous collections of possibly geo-tagged *JSON* objects, in such a way they can write complex queries by means of high-level operators. Section 3 introduces the *J-CO* framework.

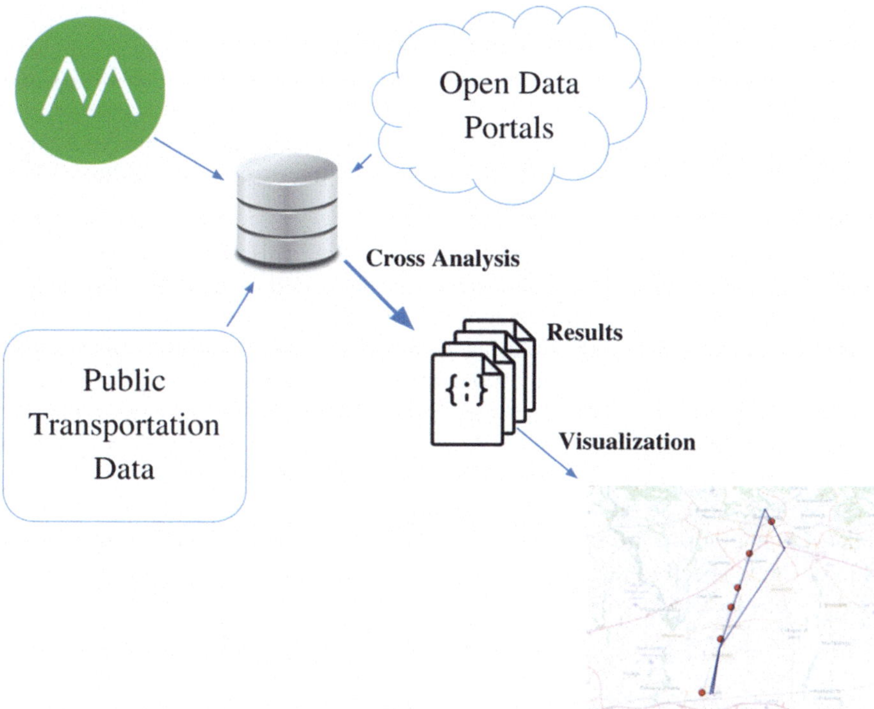

Fig. 1 Visual description for our project task: find people traveling by public transport

2.3 The Participatory Process

In the first steps of the project, we were focused on the identification of the approach to follow to start the *VGI* campaign. In particular, the issues to address were manifold:

- First of all, we had to choose the category of city users we wanted to focus on: we decided to consider high school and university students going to Bergamo each morning, in order to understand problems related to commuting and to the availability of public transportations.
- Second, once chosen the *Moves* app, we had to build a registration system that allowed new *Moves* users to register to our project; an off-line tool downloads traces of registered users.
- Third, we had to organize the participatory process and to promote it to an initial pool of students.

The first characteristic of this process is that a volunteer must be interested and passionate about the project, as he/she has to perform several tasks, in order to participate:

- Download the *Moves* app, from the app store.
- Register to the app and get the registration code.
- Visit the project page, select the registration link and provide the *Moves* registration code.
- Check if the *Moves* app is actually running (as sometimes, it stops recording).
- Through the *Moves* Web site, access his/her traces and, if he/she wishes, annotate traces and places he/she visited.

As the reader can see, the process is not so easy. This is due to the fact that *Moves APIs* are structured in such a way they do not allow new users to register from external applications. Furthermore, the app showed to be strangely unstable.

The campaign was presented during focus groups and about 500 students participated to these meetings. About 100 students registered to the project page and, currently, only 30 students are actively tracing themselves. This means that the production of *VGI* must be motivated and the active role of city users must be continuously strengthened in order to involve them in the long term.

Nevertheless, in the framework of the spatial analysis, the sample number of 30 students is enough in order to produce a deeper analysis of their movements, by asking them to specify the means of transport, the reasons of the movement and the typology of places they visited.

3 J-CO Framework

In this section, we will introduce and describe the *J-CO* framework (which stands for *J*SON *CO*llections) that has been devised to provide non-programmers with a powerful data management framework for VGI and multi-source information cross-analysis. Main characteristics of this framework should be the ability of storing and querying collections of possibly geo-tagged *JSON* objects, performing at the same time complex spatial operations; furthermore, users should be provided with tools to visualize data within classical *GIS* environments.

Figure 2 depicts the high-level components of the *J-CO* framework: hereafter, we describe the main features of the framework.

- The storage service is provided by *MongoDB*, the popular *NoSQL* database system, which is able to store collections of heterogeneous *JSON* objects. *MongoDB* is a *NoSQL* database, suitable to store *JSON* data-sets (the reader can refer to [11–14] for helpful introductions to the NoSQL database world).
- The framework is founded on a *Data Model* and an *Execution Model* that make possible to define a *query language* which we call *J-CO-QL*. This novel query language has been specifically designed to specify complex geo-spatial analysis tasks on heterogeneous collections of *JSON* objects.

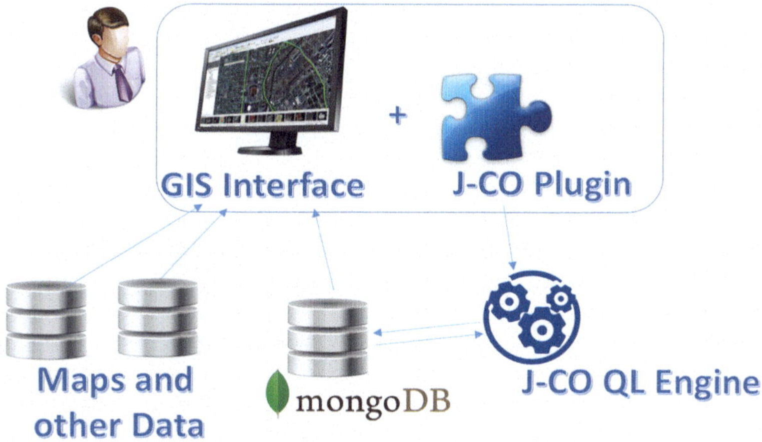

Fig. 2 Components of the J-CO-QL framework

- The *J-CO-QL Engine* executes queries on the collections stored within *MongoDB* databases and stores results into (possibly different) *MongoDB* databases.
- The *J-CO-QGIS Plug-in* provides *QGIS* (the free and widely used *GIS* tool) users with the possibility to explore collections in *MongoDB* databases and geo-tagged *JSON* objects and their geometries as spatial layers, so that they can be overlaid with other information layers.

3.1 The J-CO-QGIS Plug-in for QGIS

In order to provide analysts with a powerful tool for querying and visualizing data within classical *GIS* software, the *J-CO* framework encompasses a plug-in for *QGIS*, the free *GIS* software.

The plug-in, named *J-CO-QGIS,* provides useful functionalities. In particular:

- A *DB Browser*, that allows users to connect to *MongoDB* persistent databases to select the collections to show.
- A *Collection Viewer*, that takes collections from *MongoDB* persistent databases and loads those objects with geometrical representation into *QGIS*.
- A *Query Issuer*, that is, a text editor that allows users to write *J-CO-QL* queries and send them to the *J-CO-QL* engine.

The plug-in adds two buttons to the *QGIS* tool-bar: the first one opens the *DB Browser* window, while the second one opens the *Query Issuer* window, where a user can run *J-CO-QL* queries.

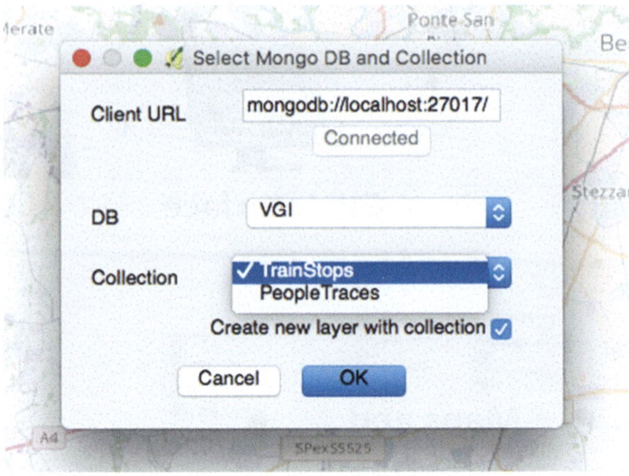

Fig. 3 *J-CO-QGIS* plug-in's *Db Browser*

3.1.1 DB Browser

Figure 3 shows the *DB Browser*: the user specifies the connection string to the *MongoDB* server. By clicking on the *Connect* button, the plug-in actually connects to the specified *MongoDB* server and shows the list of available databases. By browsing this list, shown in the *DB* list-box, the user chooses the database from which to get the desired collection. The content of the *Collection* list-box is updated with the list of collections available in the chosen database.

In Fig. 3, we connected to our *MongoDB* server, and selected the `VGI` database, that contains two collections: *TrainStops* and *PeopleTraces*. We selected the first collection.

The selected *Create new layer* functionality with *collection* flag selected causes the immediate generation of a *QGIS* layer with all geo-tagged objects contained in the chosen collection. After pressing the *OK* button, the *Collection Viewer* window is open.

3.1.2 Collection Viewer

In this window, it is possible to select objects of interest to add to the layer; the right-hand side area shows the full structure of selected *JSON* objects.

Suppose that a complex process has been performed, obtaining a new collection named `PeopleTraces`, that describes traces of people. If we select objects related to traces of some people of interest, we can easily overlay this new layer with the one created for collection `TrainStops`, as shown in Fig. 4.

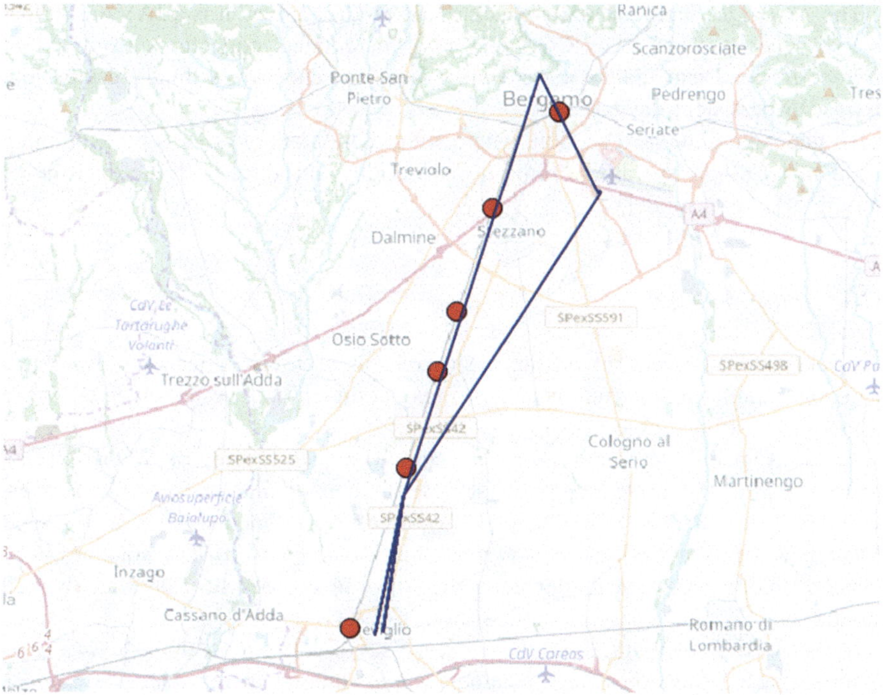

Fig. 4 Combined visualization of stations (red bullets) and of a *Moves* traces (blue line) in *QGIS*

3.2 *J-CO-QL Query Language and Its Execution Engine*

The key components of the *J-CO* framework are the query language, named *J-CO-QL*, and its execution engine. The language has been designed to provide non-programmers with a high-level language, able to deal with heterogeneous objects at the same time, performing complex spatial operations.

In particular, this capability of dealing with heterogeneous objects at the same time, simplify the query process: typically, several queries must be written and then their results must be united together (as, for example, in SQL++ [15], a very interisting proposal for querying *JSON* objects). In contrast, *J-CO-QL* provides operators specifically designed to deal with objects with different structure within the same operator application.

Furthermore, *J-CO-QL* directly deals with spatial geo-tagging of *JSON* objects, while other languages for querying *JSON* collections completely miss support for managing spatial representation. When we designed the *J-CO-QL* language, we moved from our previous experience in query languages for collections of heterogeneous objects (see [16] and [17]).

The detailed introduction of *J-CO-QL* is outside the scope of this chapter. The reader can refer to our internal report (see [18]) for a detailed description of the language.

The *J-CO-QL Engine* actually executes *J-CO-QL* queries. It is a JAVA tool that receives queries from the *J-CO-QGIS plug-in*, extracts collections from within *MongoDB* persistent database, executes queries, and saves final results into *MongoDB* persistent databases.

To understand the next section, where we present a query process by explaining the main *J-CO-QL* operators on-the-fly, we briefly introduce both the data model and the execution model.

3.2.1 Data Model

The data model is founded on the basic concept of *JSON* object. Fields (object properties) can be simple (numbers or strings), complex (i.e., nested objects), arrays (of numbers, strings, objects).

As far as spatial representation is concerned (i.e., geo-tagging of *JSON* objects), we rely on the *GeoJSON* standard (see [19]). In particular, we assume that the geometry is described by a field named `~geometry`, defined either as a *Geometry* object type or as a *GeometryCollection* object type in *GeoJSON* standard. The absence of this top-level field means that the object does not have an explicit geo-tag.

In the *J-CO* framework, a *database db* is a set of collections: $\text{db} = \{c_1, \ldots, c_n\}$. In turns, each collection c has a name *c.name* (unique in the database) and an instance *Instance(c)* = $[o_1, \ldots, o_m]$, that is a vector of *JSON* objects o_i.

3.2.2 Execution Model

Queries will transform collections stored in databases managed by *MongoDB* and will generate new collections that will be stored again into *MongoDB* databases, for persistence. For simplicity, we call such databases as *Persistent Databases.*

The basic concept of the execution model is the concept of *Query Process State*. A *state* s of a query process is a pair $s = (tc, IR)$, where *tc* is a collection named *Temporary Collection*, while *IR* is a database named *Intermediate Results database*.

Thus, the application of an operator transforms the current state s into a new state s'. The idea is that the operator possibly takes the input temporary collection, works on it, and generates a new temporary collection. Alternatively, the operator can take a collection from the intermediate database *IR* or from a persistent database, as well as the operator can save new collections into *IR* or into a persistent database. The next operator in the query will work on the new state s', and so on. Thus, the query is simply a sequence of operator applications.

Being coherent with this execution model, *J-CO-QL* provides two families of operators: *start operators* starts the computation, taking collections from the persistent databases; *carry on operators* carry on the process continuously transforming the temporary collection and possibly saving it into the persistent databases or, for temporary results, into the intermediate result database *IR*.

The execution model provides the innovative feature of the intermediate result database *IR*. It is motivated by the need to support complex transformation processes that typically proceed through the computation of several intermediate results. In fact, it would be inappropriate to store them into the persistent databases that should store source and target collections. Furthermore, by means of *IR*, intermediate collections are clearly stated to be intermediate and disappear from the system at the end of the process. Nevertheless, *IR* is implicitly related to each single execution process, than it is automatically managed in isolation, w.r.t. other execution processes.

Finally, notice that the procedural flavor is the result of the applications of operators: they are declarative (see next section) but their application defines a process. Anyway, queries are not programs, in the sense of procedural programming languages; in contrast, queries are data transformation processes.

4 Discovering City Users Traveling by Train

As stated in the previous sections, one of the tasks for the *Urban Nexus* project is to find people who move to and from the city of *Bergamo* by public transportation. Below we introduce a series of queries in order to exemplify how this task can be conducted by means of *J-CO-QL*. This process is a simplified version of the one reported in [17], where the reader can find a more detailed introduction of the operators.

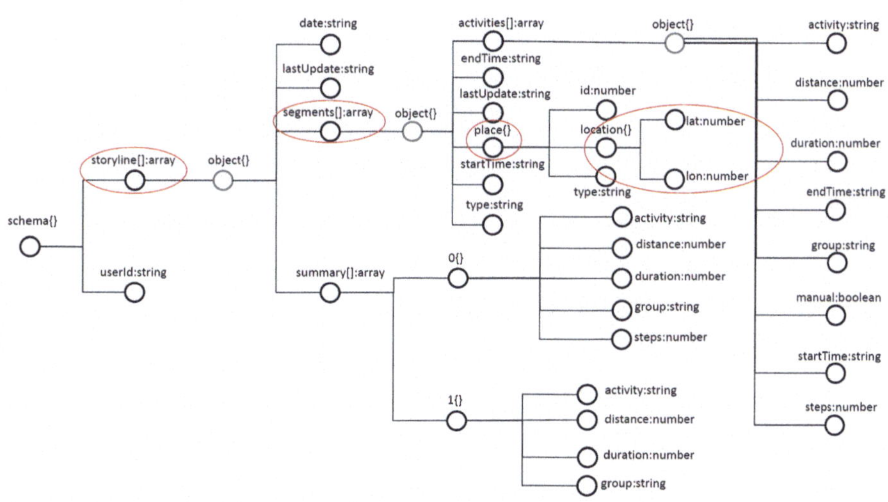

Fig. 5 Schema of *JSON* data provided by *Moves* API

4.1 Transforming Moves Data

Figure 5 shows the *JSON* schema for the data returned by the *Moves* developers' service; note that this is the schema for the data of a single user; also, note that we just need the information stored in the fields circled in red. Therefore, we need a simple way to extract such data for each *Moves* user who registered through our portal. A simple way to do that is by using the *J-CO* framework, in particular we can execute a *J-CO-QL* query that extracts all the data we need and store it in a new *MongoDB* collection.

The *J-CO-QL* query reported in Fig. 6 performs this transformation and generates collection `PeopleTraces` reported in Fig. 7 (right side).

The `GET COLLECTION` operator takes the input collection `MovesTraces` from the persistent database named `UrbanNexus`. Then a first application of the `EXPAND` operator is necessary, in order to unnest objects within the array field `storyLine` into the new field `SLItem`. That is, this new field is an object with two fields, named `item` and `position`: the first one contains the actual item extracted from the array, while the second one is the position occupied by the item.

Since `SLItem.item` is an object that contains an array of segments named segments, a second application of the `EXPAND` operator is necessary.

```
GET COLLECTION MovesTraces@UrbanNexus;
EXPAND UNPACK WITH .userId, .storyLine
      ARRAY .StoryLine TO SLItem
   DROP OTHERS;
EXPAND UNPACK WITH .UserId, .SLItem
   ARRAY .SLItem.item.segments TO seg
  DROP OTHERS;
FILTER
   CASE WHERE WITH .UserId, .Seg AND .Seg.item.type = "place"
     GENERATE {.UserId, Date: .SLItem..item.Date,
        startTime: .seg.item.startTime,
        endTime:          .seg.item.endTime,          location:
.seg.item.location}
     SETTING GEOMETRY POINT(.seg.item.place.location.lat,
                            .seg.item.place.location.lon)
   DROP OHERS;
SAVE AS PeopleSinglePlaces@UrbanNexus;
GROUP PARTITION WITH .UserId, .Date, .startTime, .endTime,
                     .location, .~geometry
     BY .userId, .date INTO .Trace SORTED BY .startTime
     GENERATE SETTING GEOMETRY To_POLYLINE( .Trace )
   DROP OTHERS;
SAVE AS PeopleTraces@UrbanNexus;
```

Fig. 6 Process "Transforming Data concerning Volunteers"

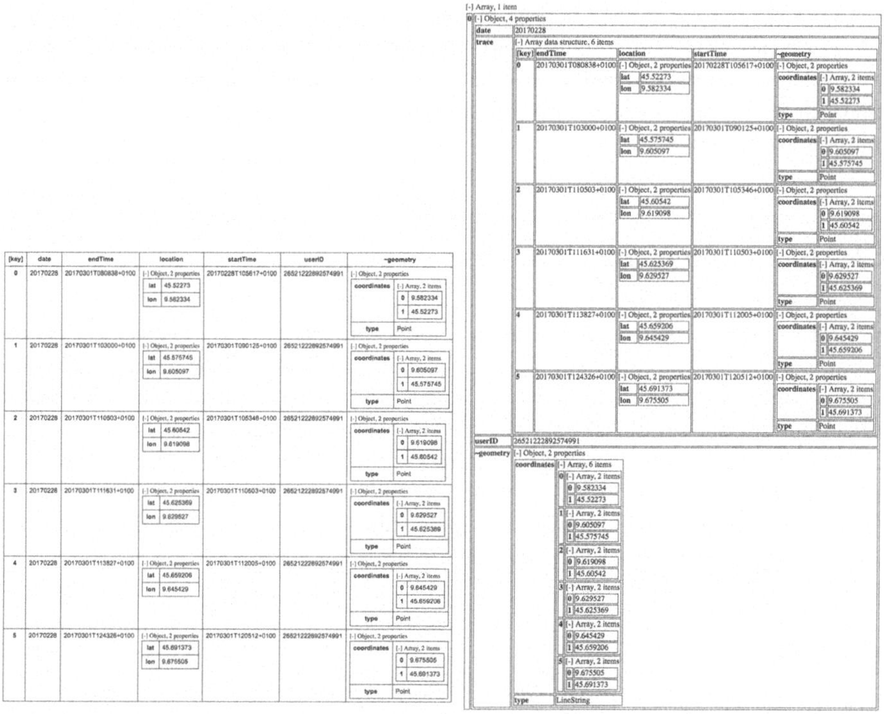

Fig. 7 Intermediate collection (left) `PeopleSinglePlaces` (set of all places visited by people) and (right) collection `PeopleTraces` (each user has, for each date, the trace within a field)

The result is a temporary collection of objects, where each object describes, in the `seg` field, each segment originally contained in the segments field. Some of them have field `.Seg.item.type` with value "`place`", other with value "`move`": we want objects with type "`place`".

To this purpose, the `FILTER` operator is used: the `WHERE` condition selects the desired objects and transforms them into simplest objects, deriving the geometry from `lat` and `lon` fields nested within the complex field `place`.

The resulting collection is saved into the persistent database `UrbanNexus` with name `PeopleSinglePlaces`, which is shown in Fig. 7 (left side).

To complete the preparation of volunteers' traces, in order to visualize them in *QGIS*, we further derive collection `PeopleTraces` shown in Fig. 7 (right side). This task is performed by means of the `GROUP` operator. Objects are grouped together on the basis of fields `userId` and `date`; the array containing all grouped objects is named `Trace` and it is sorted by field `startTime`; finally, the overall geometry is derived by aggregating geometries of all points in array `Trace` into a *polyline*.

The query appears not to be immediate. This is true, but it is also true that the task to be performed is complex and in order to execute it by means of current approaches it requires to write procedures in the form of programs, while by our approach it can be solved simply by applying four high-level operators. No programs must be written.

4.2 *Discovering Train Travelers*

We are now ready to discover train travelers by using collection `PeopleSinglePlaces` (obtained in the previous step) and collection `LineStopsStations`, obtained by outer sources of information. This latter collection describes train stations in train lines; it is depicted in Fig. 8.

What strategy could we apply to detect train voyagers? A train voyager is a volunteer whose trace is mostly overlapped with a train line. The *J-CO-QL* query reported in Fig. 9 shows the way in which train voyagers could be detected. The key of the process is the *spatial join* operation.

Fig. 8 Collection `LineStopsStations`, which describes train lines and stations

[key]	line	order	province	stationName	~geometry
0	Bergamo - Treviglio	1	Bergamo	Bergamo	[-] Object, 2 properties **coordinates** [-] Array, 2 items **0** 9.675082 **1** 45.690271 **type** Point
1	Bergamo - Treviglio	2	Bergamo	Stezzano	[-] Object, 2 properties **coordinates** [-] Array, 2 items **0** 9.644962 **1** 45.658776 **type** Point
2	Bergamo - Treviglio	3	Bergamo	Levate	[-] Object, 2 properties **coordinates** [-] Array, 2 items **0** 9.628753 **1** 45.625131 **type** Point
3	Bergamo - Treviglio	4	Bergamo	Verdello	[-] Object, 2 properties **coordinates** [-] Array, 2 items **0** 9.619797 **1** 45.605715 **type** Point
4	Bergamo - Treviglio	5	Bergamo	Arcene	[-] Object, 2 properties **coordinates** [-] Array, 2 items **0** 9.605444 **1** 45.57451 **type** Point
5	Bergamo - Treviglio	6	Bergamo	Treviglio	[-] Object, 2 properties **coordinates** [-] Array, 2 items **0** 9.580281 **1** 45.521762 **type** Point

```
SPATIAL JOIN OF COLLECTIONS
  PeopleSinglePlaces@UrbanNexus AS Trace,
  LineStopStations@UrbanNexus AS Stop
 ON DISTANCE(M) <= 500 SET GEOMETRY LEFT
 CASE WHERE WITH .Trace.userId, .Trace.Date, .Trace.startTime
       AND WITH .Stop.line, .Stop.order,
             .Stop.stationName, .Stop.province
 GENERATE {UserId: .Trace.userId, Date: .Trace.Date,
       startTime: .Trace.startTime,  line: .Stop.line,
       stationName: .Stop.stationName,
       province: .Stop.province}
 KEEPING GEOMETRY
 DROP OTHERS;

GROUP
   PARTITION WITH .userId, .Date, .line
   BY .userId, .Date, .line
   INTO .list
   GENERATE {.userId, .date, .line, .list }
   DROP OTHERS;

SET INTERMEDIATE AS PeopleByTrain;

JOIN OF COLLECTIONS PeopleByTrain AS PBT,
                    PeopleTraces@UrbanNexus AS T
  CASE WHERE .PBT.userId = T.userId AND .PBT.date = T.Date
             AND SIZE(.list)>=2
  GENERATE {userId: .T.userId,
             date: .T.date,
             trace: .T.trace}
  SETTING GEOMETRY .T.~geometry
  DROP OTHERS;
SAVE AS PeopleByTrain@VGI;
```

Fig. 9 Process "Discovering Train Travelers"

The SPATIAL JOIN operator at the beginning of the query is used to find out points in volunteers' traces that correspond to train stations: the spatial join is performed on objects in collection PeopleSinglePlaces (depicted in Fig. 7) and collection LineStopsStations (depicted in Fig. 8); they are, respectively, aliased as Trace and Stop.

The spatial join occurs if the trace point is less than 500 m from the station point. For each pair of spatially joined objects, a new output object is generated (see the GENERATE action), which reports the user id, the date, the time, the train line, the order in the line, the station name, and the province.

The collection is saved into the intermediate results database *IR* with name `PeopleInStations`.

In a heuristic way, we can make the hypothesis that a user that is found close to at least two different stations of the same line in the same day, actually traveled by train. The `GROUP` operator groups items resulting from the previous spatial join, generating an object for each user in a given date and on a given line; resulting objects contain an array named `list`, that contains all grouped objects. The resulting collection is stored into the intermediate result database with name `PeopleByTrain`.

The last `JOIN` operator in the query is necessary to produce the final collection, then saved into the persistent database `UrbanNexus` with name `PeopleByTrain`, that contains traces of people discovered in the previous steps. It is the subset of traces in collection `PeopleTraces` that actually reveals people traveling by train, adding the geometry of the traces, so as to be able to show those traces in the QGIS environment. This work is performed by the `JOIN` operator, that similarly to the `SPATIAL JOIN` operator, couples objects coming from the two collections; the `CASE WHERE` clause specifies the actual join condition.

Once people traveling by train are discovered, the analyst can conduct an exploratory activity through the *J-CO-QGIS* plug-in.

Again, notice the relative simplicity of the query, that permits to perform such a complex task with a limited number of operators. The reader could think that the large number of clauses can be difficult to understand; however, the high-level of the operators allows non-programmers, yet skilled data science researchers, to understand how to use the operators and effectively apply them to their data analysis tasks.

5 Conclusions

The goal of this chapter is to show how a multidisciplinary project can foster the development of new computer science solutions for data science activities by means of a strong collaboration with geographers and spatial analysts. Namely, the *Urban Nexus* project is devoted to exploit big data to the analysis of urban mobility, in order to understand how inhabitants and city users live the city.

Since several sources of information must be exploited, in particular institutional data, open data from public administration portals and VGI, tools that allow non-programmers (in our project, geographers) to easily analyze data are needed.

The result of our research, as far as the computer science side of the project is concerned, is a data management framework, named *J-CO*, that stores heterogeneous collections of possibly geo-tagged data into a NoSQL database, namely *MongoDB*. The framework supplies a query language (named *J-CO-QL*) that provides users with a powerful mean to deal with heterogeneous objects in collections and performs complex spatial analysis in a straightforward way; finally, but not least, the framework provides a plug-in for *QGIS*, in order to directly issue queries from and visualize data in the *QGIS* environment. From the spatial analysis side, once we use the *J-CO* framework, we can start making a cross-analysis and mapping in order to better understand data produced by the students (our sample set of volunteers for the analysis) and better understand the critical points of the public transportation system.

We think that this is a very exciting experience: from the cooperation of researchers belonging to very different disciplines (geography, statistics, computer science) point of view, our project has demonstrated that, by combining a citizen science approach to help the collection of data and the organization of participatory processes to collect *VGI* and a database approach to help data manipulation and analysis, innovative solutions to data science problems can be provided, so that both data science researchers and geographers are provided with high-level tools that foster their research work.

References

1. Lévy, J. (2003). Capital spatial. In J. Lévy & M. Lussault (Eds.), *Dictionnaire de la géographie et de l'espace des sociétés* (pp. 124–126). Paris: Belin.
2. Bordogna, G., Capelli, S., & Psaila, G. (2017). A Big geo data query framework to correlate open data with social network geotagged posts. In *Proceedings in AGILE 2017 International Conference* (In press).
3. Goodchild, M. F. (2007). Citizens as sensors: The world of volunteered geography. *GeoJournal, 69*(4), 211–221.
4. Lévy, J. (Ed.). (2008). *L'invention du monde*. Une géographie de la mondialisation: Presses de Sciences Po, Paris.
5. Soja, E. (2000). *Postmetropolis: Critical studies of cities and regions*. London: Nlackwell.
6. Casti, E., & Burini, B. (Eds.). (2015). *Centrality of territories: verso la rigenerazione di Bergamo in un network europeo*. Bergamo: Bergamo University Press.
7. Graham, M., & Shelton, T. (2013). Geography and the future of big data, big data and the future of geography. *Dialogues in Human Geography, 3*(3), 255–261.
8. Lussault, M. (2007). *L'Homme spatial*. Paris: Seuil.
9. Casti, E. (2015). *Reflexive cartography: A new perspective in mapping*. Amsterdam: Elsevier.
10. Burini, F. (2016). *Cartografia partecipativa*. Franco Angeli, Milano: Mapping per la governance ambientale e urbana.
11. Cattell, R. (2011). Scalable SQL and NoSQL data stores. *SIGMOD Record, 39*(4), 12–27.
12. Han, J., Haihong, E., Le, G., & Du, J. (2011). Survey on NoSQL database. In *2011 6th International Conference on Pervasive Computing and Applications (ICPCA)* (pp. 363–366). Hoboken: IEEE.
13. Nayak, A., Poriya, A., & Poojary, D. (2013). Type of NOSQL databases and its comparison with relational databases. *International Journal of Applied Information Systems, 5*(4), 16–19.
14. Robin, H., & Jablonski, S. (2011). NoSQL evaluation: A use case oriented survey. In *CSC-2011 International Conference on Cloud and Service Computing,* Hong Kong, China, December (pp. 336–341).
15. Ong, K. W., Papakonstantinou, Y., & Vernoux, R. (2014). *The SQL++ unifying semi-structured query language, and an expressiveness benchmark of SQL-on-Hadoop, NoSQL and NewSQL databases*. CoRR, abs/1405.3631.
16. Bordogna, G., Pagani, M., & Psaila, G. (2006). Database model and algebra for complex and heterogeneous spatial entities. In *Progress in Spatial Data Handling* (pp. 79–97). Berlin: Springer.

17. Bordogna, G., Capelli, S., Ciriello, D. E., & Psaila, G. (2017). A cross-analysis framework for multi-source volunteered, crowdsourced, and authoritative geographic information: The case study of volunteered personal traces analysis against transport network data. *Geo-spatial Information Science*, 1–15.
18. Capelli, S., Fosci, P., Marini, F., & Psaila, G. (2017). *J-CO-QL: A flexible query language for complex geographical analysis of heterogeneous geo-tagged JSON data sets*. Technical Report. http://cs.unibg.it/psaila/j-co-ql/internal-report.pdf.
19. Butler, H., Daly, M., Doyle, A., Gillies, S., Hagen, S., & Schaub, T. (2016). *The GeoJSON Format*. Technical Report.
20. Psaila, G. (2011). A database model for heterogeneous spatial collections: Definition and algebra. In *2011 International Conference on Data and Knowledge Engineering (ICDKE)* (pp. 30–35). Hoboken: IEEE.

Part III
VGI Quality and its Management and Assessment

A Chimera of VGI, Citizen Science and Mobile Devices

Vyron Antoniou

Abstract Volunteered geographic information (VGI) eloquently sums a wide variety of activities that generate geospatial content. While this content has taken many forms this chapter focuses on data sets that are used in the context of geographical citizen science projects. Citizen science covers a wide range of activities which in essence empowers communities to act on available or newly gathered data. However, there are issues of concern when it comes to the volunteers' capacity to support a sustainable flow of appropriate data. In this context, this chapter immerses into one of the major factors that play a crucial role in the development of geographical citizen science: data quality. We first discuss the characteristics of VGI and then use this experience on citizen science project that has as a basic component the collection of geographic information. We present the quality challenges that such projects can face, and we discuss in a step-by-step mode best practices that can enhance data quality.

1 Introduction

In Greek mythology, Chimera was a creature composed of the parts of three animals: a lion's head, a goat's body and a snake's tail. This incongruous combination was since used to express either anything that is seemingly composed of very disparate parts or a vain dream or idea.

In present days, we experience the convergence of two different efforts: citizen science and volunteered geographic information (VGI). Citizen science refers to scientific work that comes from the public either in collaboration or under the direction of professional scientists [34]. VGI is a term coined by Mike Goodchild, in an effort to describe '*the widespread engagement of large numbers of private citizens, often with little in the way of formal qualifications, in the creation of*

V. Antoniou (✉)
Hellenic Army General Staff, Geographic Directorate, Military Camp Papagou, Mesogeion 227-231, 15561 Holargos, Greece
e-mail: v.antoniou@ucl.ac.uk

G. Bordogna and P. Carrara (eds.), *Mobile Information Systems Leveraging Volunteered Geographic Information for Earth Observation*, Earth Systems Data and Models 4, https://doi.org/10.1007/978-3-319-70878-2_7

geographic information' ([17], p. 217). This convergence of efforts takes place in a relatively new, and more importantly, rapidly changing technological background: mobile devices. Citizens usually using their smartphones' sensors, cheap do-it-yourself devices or more advanced purpose-built devices, collect data associated with geographic information in the context of citizen science or VGI projects.

Thus, here we are today, facing a modern Chimera with VGI, citizen science and mobile devices composing a seemingly incongruous triangle that is realized as an unprecedented wealth of data. However, and perhaps just like the mythical creature, this might prove a powerful combination. Interestingly enough when it comes to data, power is a tangible element. One of the most useful factors to understand the power of the data at hand is to evaluate and measure quality. Thus, it is the quality of the volunteered data that will mainly determine the power of this combination, as it will reveal the potential, the strong and the weak points that will push citizen science into the future or into obsolescence.

Having this in mind and before we turn our discussion to the quality of data that have been collected from mobile devices within projects that combine VGI and citizen science, or more publically known as geographical citizen science [20], we need to briefly discuss and understand the birth, nature, development and the prospects of VGI and citizen science independently. Thus, this chapter is structured as follows: in the next section, we discuss the VGI phenomenon; once we describe the nature of VGI, we extend our discussion to the quality issues that VGI data present. Then, in Sect. 3, we review the subject of citizen science and its characteristics. In Sect. 4, we discuss geographical citizen science from a quality point of view, and we present challenges and answers on the quality front. The chapter concludes with discussion and conclusions (Sect. 5).

2 Volunteered Geographic Information

2.1 Definition, Origin and Development of VGI

The term VGI eloquently sums a wide variety of activities that generate geospatial content. An important milestone in the VGI history has been the birth of OpenStreetMap (OSM—www.openstreetmap.org) in 2004. Undoubtedly, OSM has played a very important role in the development of VGI, and still today is the prime example of crowdsourced collaboration for creating geographic information (GI). However, since then VGI has considerably evolved and now has a deep and broad agenda that ranges from explicitly contributed data to applications which aspire to create complete topographic maps of the world like OSM or applications that use geo-tagged photos trying to describe places up to implicitly generated GI through social networks. In fact, VGI today can come from diverse sources and in many flavours and forms. Examples include toponyms, GPS tracks, geo-tagged photos, synchronous microblogging, social networking sites such as Facebook, blogs, gaming spaces, sensor measurements, complete topographic maps. Interestingly enough, the variety of GI available is not the most important issue. The openness and availability of GI have surfaced the merits, value and potential of this kind of information that now GI is considered as an indispensable ingredient of many scientific domains. VGI now stands in the middle of a highly interdisciplinary area that intertwines the advances of many domains. In a sense, VGI is the mixture of social, economic and technological factors with the geospatial domain. Furthermore, VGI can be considered as the medium that allows active citizens to voluntarily record spatial features or phenomena that themselves consider important to have on a map.

2.2 VGI and Quality

Notwithstanding the diffusion of VGI in geospatial or other domains, there are still concerns mainly associated with the quality of the data collected. First, the social component of VGI in susceptible to all sorts of biases. Factors like the digital divide [18], socioeconomic factors [19], user collaboration [21], population and space popularity [2] can all affect VGI quality. Then, there is the process of data collections. VGI data sets usually have no metadata and are heterogeneous, and their development necessitates the combination of patchwork and fragmented contributions with diverse nomenclatures and typologies. Nevertheless, VGI is still GI and the accumulated knowledge on the field of spatial data quality can provide an useful staring point. Indeed, there is efficient theoretical background and extensive experience on the field of spatial data quality as described in official and well-established frameworks such as the ones provided by the ISO Technical Committee 211 (ISO/TC211—http://www.isotc211.org/).

According to ISO 9000 [23], quality is the '*degree to which a set of inherent characteristics fulfils requirements*'. Both characteristics and requirements are further explained. The former (i.e. characteristics or *quality elements*) is defined as those features of a product that can be either inherent or assigned and can be either qualitative or quantitative. The latter are defined as needs or expectations that are stated, obligatory or generally implied. Thus, the important steps are first to unequivocally define the quality elements of a spatial data set and then measure how good these elements fulfil predefined requirements of the producer or the user. If the specifications of a product and the end-user requirements are used alternately as the background to this theoretical approach, then we end up with internal or external quality, respectively. This means, for example, that a spatial data sets can have high-internal quality when adhere to the specifications set by the producer but at the same time low external quality if fails to satisfy the needs of the end-user.

This approach uses a number of quality elements and measures to evaluate data quality, which is [24]:

(i) **Completeness** which refers to the presence or absence of features, their attributes and relationships compared to the product's specification;
(ii) **Logical consistency** which refers the degree of adherence to logical rules of data structure, attribution and relationships as described in product's specifications;
(iii) **Positional accuracy** which refers to the accuracy of the position of features within a spatial reference system;
(iv) **Thematic accuracy** which refers to the accuracy of quantitative attributes and the correctness of non-quantitative attributes and of the classifications of features and their relationships;
(v) **Temporal quality** which refers to the quality of the temporal attributes and temporal relationships of features; and
(vi) **Usability** which refers to how a given data set can meet specific user requirements that cannot be described using the quality elements described above.

Interestingly, even this well-studied and extensively tested quality framework falls short when it comes to measuring the characteristics of VGI data. First, for the framework to work properly a reference data set (usually from an authoritative source) of higher quality is needed. However, such reference data is not always available not least due to licensing restrictions or high procurement costs. Then, the internal or external quality is difficult to be assessed; as usual, there are no rigid specifications for crowdsourced data and user requirements might be vague or unknown. Academic research on this field tries to develop tangible indicators of data quality that will apply better to the nature of VGI data. These indicators examine factors related to the VGI data, contributors' performance, socioeconomic indexes and demographic data which can function as proxies of the VGI quality (for more see [4]).

VGI is an evolving phenomenon. In a very short period, VGI has succeeded to cover a lot of ground and undoubtedly, even more, lies ahead. What we should not lose out of sight while we are rigorously measuring VGI quality which is the true value of the phenomenon [5].

3 Citizen Science

It is interesting to note that, today, both in parallel and in cross-cutting fashion, another phenomenon is rapidly spreading. In this case, active citizens are involved in a wide range of scientific activities in citizen science projects and in essence empower communities to act on available or newly gathered data.

3.1 Definition, Origin and Development of Citizen Science

Usually, in the literature, the first steps of citizen science are located at the eighteenth and nineteenth century [34]. The early examples of inspired civilians that offered so many things to science as few did, like Benjamin Franklin and Charles Darwin and the early citizen science projects like the Christmas Bird Count, an ongoing project where scientists and volunteering observers monitor birds and habitats every year since 1900, have paved the way to what is now a widespread public attitude towards science. However, the truth is that from the dawn of human civilization, primitive inclination towards citizen science existed. In its early forms, citizen science sprang out of inherently and intuitively curious individuals that could allocate time and afford the expenses to satisfy their curiosity. From the ancient civilizations (e.g. Egyptians, Greeks and Chinese) up to now, the inherent urge to measure, experiment and understand the world around us is present. Perhaps the fact that this is a deeply rooted human characteristic is what makes citizen science so interesting. Mathematics, meteorology, ecology and astronomy are just few domains that own their development to citizens.

In modern years, this human impulse was suppressed by the divide between official scientists and citizens. However, today the advancements in information technology (IT), the widespread use of mobile devices (equipped with multiple sensors) and the social interconnection over the Web have functioned as power multipliers in the hands of active citizens. Moreover, in the technological front, Web 2.0 provides a fertile environment for the creation of user-generated content, collaboration and collective intelligence. This is supported by the reduced cost of mobile devices which are equipped with more and more sensors including GPS receivers and the constantly increasing adoption of technological standards that enable interoperability. In the societal front, the shrinking digital divide, the growth in the population of well-educated individuals, the interconnection between citizens (not least due to the impact of social networks) and the increase in outdoor and leisure activities in economically advanced societies allow the collaboration of citizens on projects of common interest. Thus, in its current form, citizen science refers to the active participation of non-professional scientist in scientific projects. Citizen science runs through all levels of society and vice versa. This means that citizens participate in most, if not all, stages of the scientific process. Citizens can offer their workforce, resources and intellectual effort in any stage of a citizen science projects and although it is usually focused in data collection, it can expand to the analysis and dissemination of the results [7, 10, 34]. Their participation takes place in a collaborative manner alongside with professional scientists [25] which often can involve the identification of the needs, the layout of scientific projects and the definition of goals and achievements [27]. Yet, as the scientific challenges become more and more specialized and difficult, citizen science needs to prove its merit. In this context, citizen science projects need to take into account ethical and personal data issues, implement good practices and protocols and follow adopted policies. But, above all, it needs to be effective and delivers on their goals. This cannot be done without the collection and use of quality data, a prerequisite for trustworthy results.

Citizen science project can attract interest from various areas of the society. The stakeholders can be academic and research institutions, scientists and researchers employed in industry, the public sector and the government as well as in non-governmental organizations (NGOs). Additionally, important role plays individual volunteers, citizen organizations and associations, social enterprises and organized local or national communities. While the role of scientists is well established in scientific projects, the level of citizens' participation might considerably vary. For example, [20] presents a typology on the level of participation that ranges from '*crowdsourcing*', to '*distributed intelligence*', to '*participatory science*' and finally to '*extreme citizen science*'. By examining the goals of citizen science projects and the importance of physical environment to participation [39], group the types of volunteered participation in '*action*', '*conservation*', '*investigation*', '*virtual*' and '*education*'; while more typologies on citizens' participation can be found in [7, 11, 41].

However, irrespectively of the stakeholders and the type and level of participation, the value of citizen projects is high. Regarding society, the democratization of science, the transparency in the processes and the results of scientific projects and the engagement of the general public into projects funded by their taxes is just a few. Citizen science can help in building better societies, as it provides important educational benefits [29, 32]; it can raise awareness regarding common problems and challenges in a local and global scale, and it can increase individual capacities and skills.

On the flip side, scientist and scientific projects can also gain a lot by this engagement. For example, it is almost mandatory for any project that needs in situ data over the globe or large geographic areas, to drum up support from the public [34]. Especially for environmental projects citizens' engagement can counterbalance the deficit in up to date professional or governmental data [33]. In parallel, scientists themselves can reinforce or expand their horizons from a larger variety of ideas and more meaningful interpretations of data and results [30]. Moreover, the workforce and the computational power that the public can put in the service of scientific endeavours are immense. Examples can be found in projects like *Zooniverse* (https://www.zooniverse.org/) (the largest platform of collaborative research) or in *Foldit* (https://fold.it/) aiming to solve the protein folding problem. Thus, citizen science projects can lead to superior research results compared to traditional research project [20].

Notwithstanding all the positives that citizen science projects bear, the fact is that there is still a lot of scepticism around this domain. This is due to the nature of citizen science, enhanced by the public view that scientists know better the needed scientific principles while public participation is unfiltered and uncontrolled and thus prone to errors in data gathering and results.

3.2 Internal Characteristics of Citizen Science Projects

Before turning our focus on the quality challenges that citizen science projects face, it is necessary to briefly discuss the nature of citizen science as it will provide a better understanding of the errors that can occur and can compromise the integrity and the value of such projects.

First, we have to consider that the need to collect high volumes of data in large temporal and spatial scales makes coordination and supervision of citizen science projects a challenging task; a task that is far from what researchers are doing in a confined laboratory environment [12]. Part of this challenge is that at the beginning of citizen science projects, there is a need to devise a strategy on how to effectively engage volunteers [33]. A project needs to successfully drum up support and develop lasting public engagement which will not be prone to biases (e.g. due to race, gender, socioeconomic status or other factors). Despite this, internal biases are hard to escape. Usually, the educational and technical skills are needed; the access to resources and the free time correlate with individuals that live in advanced economies [20]. In turn, this might lead to a form of 'elitism' that excludes or drives out individuals that cannot allocate the resources need to match the level of participation of other volunteers. Next is the spatial coverage: as is the case with VGI projects, spatial biases are common since high populated and touristic areas are expected to attract participants while more obscure and hard to reach areas will be under-represented in the data collection phase. Spatial biases are coupled with temporal ones. Working hours and days are less likely to attract participation compared to after-work hours, weekends and holidays, especially when the project needs in situ data gathering. In a next level, volunteers' input might affect the decisions on methods, processes and analyses followed in the course of an experiment or project. Thus, only individuals that have in-depth training and can be of help should be able to contribute in these phases of the project. Consequently, the onus is on the scientists that need to provide concrete answers regarding the setup of the project and the monitoring and management of volunteers' participation so to avoid unwanted biases. All these lead to concerns about lower quality of data that has been collected by volunteers or the level of appropriate data management and curation due to unknown and varying levels of knowledge and skill on behalf of volunteers [33].

On the other hand though, citizen science attracts the interest of self-conscious and active people that approach volunteerism as part of their way of living and thus their participation is not, at least a priori, harmful to the goals and aims of a project. A step further, an individual with basic education today, is in place to follow, understand and support the many phases of the project, with or without supervision, and thus provide quality input to the project. All these results in a constant struggle to keep the balance between these internal to citizen science characteristics and the need for quality inputs.

4 Quality of GCS: Challenges

The worlds of VGI and citizen science meet in what is termed '*Geographical Citizen Science*' (GCS) where the collection of positional information is an integral part in the course of a citizen science project [19]. As expected, this field carries all the positives and negatives from both worlds. While in many cases the merge is beneficial as, for example, many environmental projects have gained a lot from the VGI [9], the most important undermining factor to the integrity of both VGI and citizen science projects is the part where volunteers leave their biggest footprint: data quality.

4.1 Quality: Challenges

First, we need to understand the role that data quality plays in GCS projects. The huge volumes of data that can be gathered for large geographic areas can be used by policy-makers, authorities and governments to make assessments or define policies that might affect stakeholders at all levels of public life. Thus, the reliability of the data that have been used in a project is of high importance due to the potential political, economic and societal implications of the findings [33].

In a sense, the quality of the data collected mirrors the reliability and value of the project itself. In many cases, data collected in the course of a project might be reused to other studies or projects. If data has not been quality controlled imprecise and biased observations could have permeated and thus negatively affect scientific efforts. Negative reputation might cascade to other endeavours and might affect the whole citizen science movement. Moreover, sustainable user engagement, funding and the longevity of a program can be severely affected if the data gathered proved to be of inferior quality.

The fact is that, the very nature of GCS projects imposes the data collected by citizens to be heterogeneous and patchy. Attention to detail, granularity of the data and documentation of the measurements can greatly vary depending on the knowledge, level of skills and the resources and the time devoted by citizens. Moreover, the devotion of the public could vary as well. As [16] note, citizen science data sets will include measurements from wherever volunteers happen to report them and that the gathering processes can very well suffer from volunteer boredom. Moreover, as [34] notes there is a phase where volunteers are, inescapably, learning by doing and thus their early contribution might suffer from inconsistencies and various errors. It should be expected that such collaborative projects will have contributions from large, diverse and more or less unknown pool of volunteers with variable aims and goals [40].

All these directly affect quality and thus should not all data be treated in the same way. Moreover, the scope of a citizen science project might give other perspective to the quality of the data gathered. For example, in primarily educational citizen science projects, the actual data quality might be deliberately overlooked in favour of the educational goals. Thus, the notion of data quality should be considered in the context of the GCS, its broader domain and the purpose of the data [33]. Information about quality should not be assigned to large data sets unless prior actions of quality control and harmonization have taken place. Differently, information about the quality of data should reach down to observation level.

An important factor in the discussion about quality is quality requirements. As not all geospatial products need the same level data quality, similarly various citizen science projects can find merit in data sets of varying quality. For example, not all environmental and meteorological projects need the same level of spatial density or positional accuracy in their data. Thus, explicitly defining the level of quality sought can give to any interested party a clear view on the possible uses of data and the applicability of the results generated. As [32] note, this can allow interpretive approaches to accept data of lower quality which, however, suffice for their own needs. Another factor is the instruments used for the data collection. In VGI projects like OSM, data capture takes place either from GPS receivers of known accuracy or through on-screen digitization of orthorectified satellite imagery. Thus, VGI data is generated from instruments of known potentials that encapsulate a great deal of scientific knowledge. In classic scientific project, measurements and observations take place in controlled conditions usually with expensive, calibrated and well-tested instruments. However, in citizen science projects, this might not be the case. Occasional instruments, do-it-yourself constructions, mobile sensor without calibration or testing and poor instrument maintenance can all be used to measure and record a phenomenon. Consequently, all these are potential sources of errors. What is common, however, in all kinds of project is that error sources can be found not only in the instruments used for data capturing but in their entire life cycle. Particularly for crowdsourced project errors can occur due to protocols misunderstanding or during data management, transformation or migration from one system to another [33].

An interesting subject is how VGI and citizen science projects approach quality. For example, for VGI data, there seems to be a consensus on the existing measures that can be applied on data so to evaluate quality elements (e.g. positional accuracy, thematic accuracy, temporal accuracy, etc.). On the contrary, it seems that, mainly due to the variety of citizen science projects, it is difficult to propose a precise, universally meaningful definition of data quality [38] despite the fact that the notion of what is correct and what is not is intuitive to volunteers [33]. However, what seems to be common in both types of projects is the notion of quality from the different points of view: from the point of the data producer (i.e. internal quality) and the point of data users (i.e. external quality). For GI, the level of accordance to specifications dictates the level of internal quality, whereas the usability of the data by the end-users dictates the external quality. Similarly, for citizen science projects as [37] note, the internal data quality can be associated with the notions of accuracy, reliability and consistency and the external quality with timeliness, relevance and interpretability. Another difference seems to arise from the way that quality is measured. For example [33], report that, in the River Watch project (http://www.ngrrec.org/riverwatch/), consistency (internal quality element in citizen science) is preserved through standard operating procedures (SOPs) used to collect the data. The goal in the consistency front is to minimize the heterogeneity of the measurements so that data can be meaningfully compared. While this is acceptable in the context of a citizen science project, it is fundamentally different from the attitude towards, for example, thematic accuracy (internal quality element in GI) where the focus is to ensure that measurements will be assigned in a predefined nomenclature so to ensure consistency throughout the data set. This also requires different approaches for tangibly reporting data quality, for documenting errors and for evaluating the fitness for purpose of each data set.

This observation becomes even more interesting if we try to evaluate the overall data quality within a GCS project, perhaps following a holistic approach [13]. To achieve that there should be an appropriate mixture of quality elements that will equally cover both the geospatial and phenological nature of each measurement, capable to measure and unequivocally report data quality. The quality of the diverse, two-dimensional nature of such data has attracted the interest of researchers, as it has been seen that biases, inconsistencies and errors on the geospatial dimension spill over to the phenological one. For example [31], examine the geographical sampling bias in the data collected to determine priority areas for conservation. Similarly [15], investigated the sources of spatial bias in the bird records and their impact on observations. In order to avoid this disturbing event, best practices and protocols need to be developed and followed.

4.2 Quality: Protocols and Best Practices in Every Phase

The importance of high-quality data sets and the challenges that GCS projects face have been discussed in the previous section. Here the focus will turn to the discussion of applied protocols and best practices in an effort to enhance the integrity of the measurements and the overall quality of the data. As [8] highlight, citizen science projects need to ensure and demonstrate the consistently high quality of data collected so to establish their reputation and value. A way forward is the use of protocols and best practices in every step of a project, and especially when it comes to data gathering [7]. Moreover, as noted by [13], data validation should take place whenever possible. This will have an impact both on volunteers and stakeholders, as it will show the devotion to high standards. What should not prevail is the belief that the ‘average’ observation will hide inconsistencies or outliers. Instead, the project’s design should cater for maintaining high-scientific standards and rigorous data collection methods: a prerequisite for equally rigorous results [10].

Safeguarding data quality should start from the very first step of a volunteer within the project. Often volunteers need to familiarize themselves with comprehensive instructions and guides regarding the nature, aims and processes of the project. In many cases, volunteers also need to undergo a training session in between registration and participation. The training can be adjusted both according to the entry skills of volunteers, and the role they are willing to play in the life cycle of the project. To this end [16], highlight the benefits of virtual reality training for improving citizen science quality. They suggest that on top of classical methods that include instructions and supporting materials which belong to passive training and workshops that is active training [28], virtual reality training can offer the level of knowledge needed so that volunteers will be able to reach high level of data quality even from the first contribution efforts. An important factor here is that virtual reality can mimic, perhaps better than any other in-house educational approach, the challenges and processes that exist during in situ data gathering.

Moving a step further from the training phase of the project, the focus goes to the data collection methods. In order to improve data quality, various projects carefully control and redesign their data input methodologies, usually using a minimal approach. The choice to simplify the tasks required by the volunteers is adopted with the hope that this simplification will improve data quality [6]. However, this strategy might hide multiple dangers as it could downgrade the value of data collected or alienate the more enthusiastic and active of the participants [26, 14]. Furthermore, software and intuitive user interfaces can play an important role in data collection processes. Mobile devices like smartphones and tablets that can support tailored applications are more likely to help inexperienced volunteers through a successful data collection task. On the other hand, do-it-yourself sensors or sensors that are based on modular hardware/software components like Arduino (https://www.arduino.cc/) and Raspberry Pi (https://www.raspberrypi.org/) need individuals with advanced skills who are able to understand the basic principles for data gathering depending on the project. In any case, as [33] highlight, quality challenges should be addressed through technological means where possible and data collection software can greatly help towards this direction.

The next step, after the training and the software and hardware available, is paying particular attention to the way that data collection needs to be designed. As explained, GCS projects can suffer from biases and unbalanced participation patterns. Consequently, data gathering processes should be carefully designed in such a way that will provide the necessary input to cover a project's needs. The sheer fact of gathering huge volumes of data does not guarantee a reliable research outcome, especially if the project needs specific spatial representation. For example, as [34] notes, the British Trust for Ornithology is a model case of statistical design, as it uses stratified random sampling methods to select the locations where data collection is needed and then guides its volunteers in the correct places. For this latter part, i.e. the guidance of volunteers to preselected locations can greatly benefit from gamification techniques, as they provide an easy, fun and effective alternative for counterbalancing spatial participation biases [3].

Once the strategy for data gather has been set and then the next step is to provide the necessary SOPs and protocols that volunteers need to follow so to achieve the desired outcome. This step is particularly important as it will help both novice and experienced volunteers to complete the data gathering process without intentional or not, and possibly harmful to the data quality, derogations. Protocol should have been introduced at the training phase so that volunteers will have a good understanding about what is expected from them. Moreover, as [35] notes, different stakeholders with various motives and objectives are expected to challenge the fundamental mechanisms of scientific evaluation systems. Protocols and SOPs can create a protecting barrier against such tendencies and should include the documentation of the data collected (e.g. photographs, filling of forms, etc.) so to help in quality assurance.

The next step in the data gathering process is to validate the data gathered. The steps followed so far should have done a great deal of work towards preserving data integrity. However, the openness and inclusiveness of GCS projects inescapably allow room for errors. Thus, a fresh round of data quality evaluation should be in place so to first spot and then remove any possible errors. At this stage, validation processes are equipped with the knowledge that can be generated by using data from the entire pool of volunteers. This allows to develop statistical methods and automated filters that can flag any possible unlikely observations, mismatches or outliers. By putting aside all data records which have been flagged as possible errors can allow a second, human-driven, round of validation to take place just for those. This could include scientists, or heavily engaged citizens of proven quality record, to thoroughly inspect all these cases before they can be used in the analysis phase. For example, in the eBird (http://ebird.org/) project, a large pool of local reviewers is used so to evaluate flagged submissions [36].

Up to now the discussion was about quality assurance methods (i.e. processes oriented to error avoidance). Once this phase is completed, the quality control (i.e. processes that focus on the identification of errors in the final outcome) should start on the assembled data sets. This phase is equally important as the previous ones, and it can include both computational and human-led checks. The common mechanism is the use of experts that will take the task to examine randomly selected parts of the data through various methods that are applicable to each project. For example, exploratory data analysis is a very efficient method to evaluate huge volumes of spatiotemporal data (for more see [1]). On the other hand, there are limitations that should be taken into consideration. As [8] note, the time needed to follow-up with individual participants and observations might be very large, especially if any of the previous steps discussed has failed to produce data of high quality, and it is not always possible to recognize plausible but erroneous observation.

The final step in this line of action is to communicate the experience and findings to all involved parties. This will raise awareness among volunteers regarding the gaps through which errors can slip into final data and what needs to be done to avert this. This retrospective training process will further improve the skills and the work of volunteers in order to avoid the same mistakes to happen repeatedly.

All these steps and phases of quality management should be part of a general framework that needs to be the backbone of any GCS project. Recognizing the importance of quality, various quality preserving mechanisms have been suggested both from the academia (see, e.g., [40]) and citizen science associations like European Citizen Science Association (ECSA) or the Citizen Science Association (CSA) in the USA.

5 Discussion and Conclusions

This chapter immerses into one of the major factors that play crucial role in the merge of spatial information with citizen science projects: data quality. This merge forms what is known as geographical citizen science, a special form of citizen science, where geographic information and geospatial principles should be explicitly stated, as they are important in the overall development of the project.

First, we briefly explored the nature of the VGI phenomenon and the challenges around data quality. Then we explored the world and nature of doing citizen science. Special attention was paid on the internal characteristics, and how these can affect both the process of data gathering and the quality of the data. The intertwining of these two worlds (i.e. VGI and citizen science) led the discussion to the field GCS where effort was made to shed light from the data quality point of view. Although here we just scratched the surface of the challenges and remedies regarding crowdsourced data, an outline of the main issues that GCS projects face has been given. This chapter explained that the driving force behind geographical citizen science projects is the volunteers and their modes of engagement. However, while on the one hand there is an imperative need for crowdsourced data; on the other, there are issues of concern when it comes to the volunteers' capacity to support a sustainable flow of appropriate data.

Citizens' participation should not be associated with '*citziens*'*data*', which will refer to data of inferior quality from contributors that did not bother or managed to follow the needed guidelines so to provide data of adequate quality. What needs to be clear to the stakeholders is that observations, measurements and data collected in the process of a citizen science project should not mix with view, stereotypes, feelings, estimations or public opinions. This principle needs to be secured by the internal mechanisms of each project and the meticulous efforts of both scientists and active citizens. Similarly, citizen science projects should not be egocentric. If a citizen's project seriously aspires to enter the world of science, it needs to follow the rigorous processes that 'traditional' scientific project are following. This includes to provide the ability to any other scientist to replicate the findings. Thus, making *reproducible citizen science* should be high in the agenda of any citizen science project. In turn, this means that data, processes and results should be available, reusable, comparable and reproducible by anyone, either in stand-alone mode or as part of other projects. Citizen science and GCS projects can learn a lot about this from the accumulated experience in VGI projects. For example, the diffusion of crowdsourced geospatial data in studies spatial products or in government is constantly expanding (see, e.g.,[22]).

The cornerstone that can support all these is the quality of the raw data sets that need to be free of biases, inconsistencies and errors. Biases can occur due to reasons related to volunteers, space and time. Volunteers' origin, subjectivity, socioeconomic background, knowledge and skills might introduce biases that cannot be spotted easily. Additionally, poor accessibility to certain areas due to various reasons (e.g. climate, landscape, public restrictions, criminality, etc.) can also introduce biases to the data. This phenomenon can be further intensified by the spatial distribution of demographic variables such as population density, income, age. Finally, biases can be introduced due to the time periods that volunteers allocate to the project. It should be expected that, in most cases, crowdsourced participation will occur during free time which is inherently biased towards afternoons, weekends and holidays. Next, it is the inconsistencies among the data submitted, which refer to incompatibility of certain observations with the rest. While it is not always easy to unambiguously identify observations that are not plausible, statistical methods, filters, and the review by experts of in situ documentation during data collection can help towards this end. Meticulous management of the above issues can reinforce the credibility and reliability of GCS projects, leading, in a sense, to reproducible citizen science.

Then, we immersed into more specific quality challenges and examined the necessary actions that need to be taken in order to address them. As the definition of what consists a data set of high quality for the, usually, heterogeneous citizen science projects can be rather elusive, the paper escapes from the restricted notion of the spatial data quality elements that GI science uses and tries to analyse in more practical, yet holistic, way the parameters that influence the gathering, manipulation and evaluation of crowdsourced data in a GCS project. Thus, following the sequence of steps in a GCS project, we examined the processes that can greatly enhance data quality. The rigorousness and the firm implementation of protocols and SOPs were praised as absolutely necessary ingredients. However, a word of caution needs to be said here. Such a tight collar might become a detriment to the relationships between citizens and scientists or be harmful to other important aims of a project such as education or raising awareness. Citizens should be initiated into the world of doing science along with all the commitments of this process. The procedure of justifying to and familiarizing citizens with all the precautions needed at the altar of quality is a wholly new issue on its own. Perhaps this challenge needs to be addressed in a broader context. What needs to be done is to train modern societies into citizen science, its values and potentials as well as the important steps that we need to take in order to achieve scientific goals.

In this context, the future of citizen science and GCS projects seems promising. Active and informed societies should engage with scientist to provide solutions to local or global challenges. On the flip side, large-scale data gathering projects will probably need extensive help from volunteers. Therefore, citizen science is expected to play an important and prominent role in both academia and society. Active research on citizen science is needed. Despite prime examples of successful projects, still there are many blind spots that need to be adequately addressed.

Finally, what has been deliberately just briefly mentioned is the mobile devices used for data collection. Technology is such a fast-changing field that describing the state of the art today, becomes very soon obsolete. For example, today we experience dramatic changes in the data collection technologies as we are on the verge of a new reality that develops in parallel with smartphone applications with the use of drones, wearables, Internet of things and ubiquitous sensors. In the, not so far, future, the task of in situ observation and measurements with mobile devices will move to a totally new level. Citizen science and crowdsourced mentality are ideally place to exploit this dynamic.

Returning back to the Greek mythology, it might be didactic to keep in mind that despite Chimera's immense strength, one of its own powerful characteristics was used against her by Bellerophon. The mythical hero used Chimera's fire breath to melt a leaden spear inside her mouth that made Chimera to suffocate to death. The most powerful characteristic of citizen science is volunteered contributions. Overlooking the quality of their data might become the deadly element that will cause the entire citizen science movement to decay into obsolescence.

References

1. Andrienko, N., & Andrienko, G. (2006). *Exploratory analysis of spatial and temporal data: A systematic approach*. Berlin: Springer Science & Business Media.
2. Antoniou, V. (2011). User generated spatial content: An analysis of the phenomenon and its challenges for mapping agencies. Ph.D. Thesis, University College London.
3. Antoniou, V., & Schlieder, C. (2014). Participation patterns, VGI and gamification. *AGILE, 2014*, 3–6.
4. Antoniou, V., & Skopeliti, A. (2015). Measures and indicators of VGI quality: An overview. In *ISPRS Annual Photogramming Remote Sensing Spatial Information Science*, II-3/W5 (pp. 345–351).
5. Antoniou, V. (2016). Volunteered geographic information: Measuring quality, understanding the value. *GEOmedia, 1*(1).
6. Batini, C., Cappiello, C., Francalanci, C., & Maurino, A. (2009). Methodologies for data quality assessment and improvement. *ACM Computing Surveys, 41*(3), Article 16.
7. Bonney, R., Cooper, C., Dickinson, J., Kelling, S., Phillips, T., Rosenberg, K., et al. (2009). Citizen science: A developing tool for expanding science knowledge and scientific literacy. *BioScience, 59*(11), 977–984.
8. Bonter, D. N., & Cooper, C. B. (2012). Data validation in citizen science: A case study from Project FeederWatch. *Frontiers in Ecology and the Environment, 10*(6), 305–307.
9. Catlin-Groves, C. L. (2012). The citizen science landscape: From volunteers to citizen sensors and beyond. *International Journal of Zoology*.
10. Cohn, J. P. (2008). Citizen science: Can volunteers do real research? *BioScience, 58*(3), 192–197.
11. Cooper, C. B., Dickinson, J., Phillips, T., & Bonney, R. (2007). Citizen science as a tool for conservation in residential ecosystems. *Ecology and Society, 12*(2), 11. URL: http://www.ecologyandsociety.org/vol12/iss2/art11/.
12. Cooper, C. B., Hochachka, W. M., & Dhondt, A. A. (2012). The opportunities and challenges of citizen science as a tool for ecological research. *Citizen Science: Public Participation in Environmental Research, 99*.
13. Dickinson, J. L., Zuckerberg, B., & Bonter, D. N. (2010). Citizen science as an ecological research tool: Challenges and benefits. *Annual Review of Ecology Evolution and Systematics, 41*, 149–172.
14. Everleigh, A., Jennett, C., Blandford, A., Brohan, P., & Cox, A. (2014). Designing for dabblers and deterring drop-outs in citizen science. CHI 2014, 26 Apr–1 May, 2014. Toronto, ON, Canada.
15. Ferrer, X., Carrascal, L. M., Gordo, Ó., & Pino, J. (2006). Bias in avian sampling effort due to human preferences: An analysis with Catalonian birds (1900–2002). *Ardeola, 53*(2), 213–227.

16. Goldman, A., & Preece, J. (2015). *Can virtual reality improve citizen science data quality?* Online: http://hci.cs.wisc.edu/workshops/chi2015/papers/chi15-goldman.pdf.
17. Goodchild, M. (2007). Citizens as sensors: The world of volunteered geography. *GeoJournal, 69*(4), 211–221.
18. Graham, M., Hogan, B., Straumann, R., & Medhat, A. (2014). Uneven geographies of user-generated information: Patterns of increasing informational poverty. *Annals of the Association of American Geographers, 104*(4), 746–764.
19. Haklay, M. (2010). How good is volunteered geographical information? A comparative study of OpenStreetMap and Ordnance Survey datasets. *Environment and Planning B: Planning and Design, 37*(4), 682–703.
20. Haklay, M. (2013). Citizen science and volunteered geographic information: Overview and typology of participation. In *Crowdsourcing geographic knowledge* (pp. 105–122). Berlin: Springer Netherlands.
21. Haklay, M., Basiouka, S., Antoniou, V., & Ather, A. (2010). How many volunteers does it take to map an area well? The validity of linus' law to volunteered geographic information. *The Cartographic Journal, 47*(4), 315–322.
22. Haklay, M. E., Antoniou, V., Basiouka, S., Soden, R., & Mooney, P. (2014). *Crowdsourced geographic information use in government*. Washington, D.C.: World Bank.
23. International Organization for Standardization. (2005). *Quality management systems: Fundamentals and vocabulary*. International Organization for Standardization.
24. International Organization for Standardization. (2013). *Geographic information—Data quality*. International Organization for Standardization.
25. Lakshminarayanan, S. (2007). Using citizens to do science versus citizens as scientists. *Ecology and Society, 12*(2), 2.
26. Lemay, P (2007). Developing a pattern language for flow experiences in video games. In *Situated Play, Proceedings if DiGRA 2007 Conference* (pp. 449–455).
27. Luther, K., Counts, S., Stecher, K. B., Hoff, A., & Pathfinder P. J. (2009). An online collaboration environment for citizen scientists. In *CHI '09* (pp. 239–248). New York, NY, USA: ACM.
28. Nadkarni, N. M., & Stevenson, R. D. (2009). Symposium 9: Linking scientists with nontraditional public audiences to enhance ecological thought. *The Bulletin Ecological of the Society of America, 90,* 134–137.
29. Nicholson, E., Ryan, J., & Hodgkins, D. (2002). Community data-where does the value lie? Assessing confidence limits of community collected water quality data. *Water Science and Technology, 45*(11), 193–200.
30. Nielsen, M. (2012). *Reinventing discovery: The new era of networked science*. Princeton: Princeton University Press.
31. Reddy, S., & Dávalos, L. M. (2003). Geographical sampling bias and its implications for conservation priorities in Africa. *Journal of Biogeography, 30*(11), 1719–1727.
32. Savan, B., Morgan, A. J., & Gore, C. (2003). Volunteer environmental monitoring and the role of the universities: The case of Citizens' Environment Watch. *Environmental Management, 31*(5), 0561–0568.
33. Sheppard, S. A., & Terveen, L. (2011). Quality is a verb: The operationalization of data quality in a citizen science community. In *Proceedings of the 7th International Symposium on Wikis and Open Collaboration* (pp. 29–38). New York: ACM.
34. Silvertown, J. (2009). A new dawn for citizen science. *Trends in Ecology & Evolution, 24*(9), 467–471.
35. Socientaze, (2013). *Green paper on citizen science*. Online at: http://ec.europa.eu/information_society/newsroom/cf/dae/document.cfm?doc_id=4121.
36. Sullivan, B. L., Wood, C. L., Iliff, M. J., Bonney, R. E., Fink, D., & Kelling, S. (2009). eBird: A citizen-based bird observation network in the biological sciences. *Biological Conservation, 142*(10), 2282–2292.
37. Wand, Y., & Wang, R. Y. (1996). Anchoring data quality dimensions in ontological foundations. *Communications of the ACM, 39*(11), 86–95.

38. Wang, R. Y., Storey, V. C., & Firth, C. P. (1995). A framework for analysis of data quality research. *IEEE Transactions on Knowledge and Data Engineering, 7*(4), 623–640.
39. Wiggins, A., & Crowston, K. (2011). From conservation to crowdsourcing: A typology of citizen science. In *2011 44th Hawaii International Conference on System Sciences (HICSS)* (pp. 1–10). Hoboken: IEEE.
40. Wiggins, A., Newman, G., Stevenson, R. D., & Crowston, K. (2011). Mechanisms for data quality and validation in citizen science. In *Proceedings of Workshops at the Seventh International Conference on eScience* (pp. 14–19). Hoboken: IEEE.
41. Wilderman, C. C. (2007). Models of community science: Design lessons from the field. In C. McEver, R. Bonney, J. Dickinson, S. Kelling, K. Rosenberg, & J. L. Shirk (Eds.), *Citizen Science Toolkit Conference*. Ithaca, NY: Cornell Laboratory of Ornithology.

Volunteered Metadata, and Metadata on VGI: Challenges and Current Practices

Lucy Bastin, Sven Schade and Peter Mooney

Abstract In the face of an exploding range of volunteered data initiatives, it is important to maintain good metadata and quality information in order to ensure the appropriate combination and reuse of the resulting data sets. At the same time, there is increasing evidence that validation and quality assessment of data (whether that data be volunteered or 'official') can sometimes be usefully crowdsourced, i.e. the required efforts can be distributed to a large number of people. However, as with VGI itself, maintaining the consistency, semantics and reliability of volunteered metadata presents a number of challenges. Initiatives which archive the history of features and tags (e.g. OpenStreetMap) lend themselves to some mapping of disputed features, but among citizen science projects in general there is often limited scope for users to comment on their own or others' submissions in a consistent way which may be translated to any of the currently accepted geospatial metadata standards. At the same time, platforms which allow the publication of more 'authoritative' data sets, (e.g. Geonode and ArcGIS Online), have introduced the option of user comments and ratings. Volunteered metadata (on both authoritative and VG information) is potentially of huge value in assessing fitness for purpose, but some form of standardization is required in order to aggregate diverse 'opinions' on the content and quality of data sets and extracts the maximum value from this potentially vital resource. We discuss major challenges and present a set of examples of current practice which may assist in this aggregation.

L. Bastin (✉) · S. Schade
European Commission, Joint Research Centre (JRC),
Via E. Fermi 2749, 21027 Ispra, VA, Italy
e-mail: l.bastin@aston.ac.uk

L. Bastin
School of Engineering and Applied Science,
Aston University, Birmingham B4 7ET, UK

P. Mooney
Department of Computer Science, Maynooth University,
Maynooth, Co. Kildare W23 F2H6, Ireland

G. Bordogna and P. Carrara (eds.), *Mobile Information Systems Leveraging Volunteered Geographic Information for Earth Observation*, Earth Systems Data and Models 4, https://doi.org/10.1007/978-3-319-70878-2_8

1 Introduction

Quality assurance is a topic as old as data collection itself. A detailed categorisation of the different dimensions of data quality is provided in [1], and these facets are related to VGI specifically in [2]. The validation of data, especially of collections of inputs using smartphones, turns out to be one of the most challenging and challenged exercises in relation to VGI data quality. This complexity relates not only to the variety of possible validation methods but also to the trust of user communities in different mechanisms. In the remainder of this section, we discuss the different kinds of metadata and then examine the value of metadata for evaluating fitness for purpose, highlighting some of the ways in which VGI can be especially challenging and poorly documented. Since one proposed solution is to crowdsource the documentation of resources, we review both experimental and more mature approaches to this activity. In Sect. 2, we explore the stages of data production and use at which metadata can be generated, considering both metadata specific to VGI and volunteered metadata which may apply to 'authoritative' or to volunteered data sets. Throughout this section, we give real-world examples of some tools, standards and approaches which are currently in use across various domains, taking some insights and inspirations from the world of non-geographic data such as cultural heritage collections. In Sect. 3, we document two specific case studies (OpenStreetMap and an invasive species monitoring app) with reference to the previously defined stages and facets of the metadata production flow. Overall, we identify and summarise ongoing challenges which must be surmounted to maximise the value and reusability of VGI and to glean the maximum knowledge from users on the data sets that they discover and use. Our conclusions are presented in Sect. 4.

1.1 Metadata and Its Uses

Lawrence et al. [3] distinguish between five (not completely exclusive) types of metadata, and this distinction is of use for the discussion at hand. Blower et al. [4] discuss this taxonomy in the context of climate data and note the key distinction between *intrinsic* metadata and *extrinsic* metadata. Intrinsic metadata (Archive, Browse, Discovery and Extra) is, in general, generated by the data producer, while character metadata, (extended to 'Commentary' metadata by Blower et al. *op. cit.*) is extrinsic to the actual data set, can be added after publication in the form of an annotation, and is often highly reflective of the domains within which the data set has been used. In this chapter, we discuss all five types of metadata as they relate to VGI, with particular references to the stages of production and use at which they may be attached to the data set, and the available tools and standards which can assist in the process.

In brief, Archive (A) metadata is the detailed description of units, spatial referencing, etc., without which the data cannot be properly used. Browse (B) metadata describes the context of a data set's production; it is usually standardised in some way and can be used (in combination with Discovery (D) metadata, which records overarching data set features such as extent and resolution for use in catalogues) to assess fitness for purpose and choose between alternative data sets or evaluate the appropriateness of combining data. Commentary (C) metadata consists of items such as usage reports, citations, reviews and quality assertions. Extra (E) metadata is defined as very structured descriptions of sensors or protocols which are generally very discipline-specific.

Existing community standards for geospatial metadata (for example, ISO 19115[1]/19157[2] or the CSDGM standard[3] of the Federal Geographic Data Committee) contain elements capable of encoding and communicating most facets of the Archive, Browse and Discovery metadata. Some elements of Commentary and Extra metadata can also be recorded using these models and encodings, and, given the substantial community efforts which have gone into implementing and profiling these standards (e.g. for INSPIRE) and developing catalogues that efficiently exploit and crosswalk them, it is very much worth aligning efforts in VGI metadata with existing standards where possible. However, not everything that is needed to evaluate VGI can be captured using these standards; commentaries, reviews and usage reports are often collated using more lightweight annotation formats, while recording technical information on protocols and tools may require the use of domain-specific ontologies, or technical and mathematical encodings such as SensorML.[4] It is unlikely and, indeed, undesirable that the VGI and classical geospatial worlds will converge on a single metadata standard: the important challenge is to reuse the best of what exists and facilitate translation, harmonisation, aggregation and enrichment of any metadata which is available.

1.2 How Metadata Helps in Assessing VGI Quality and Fitness for Purpose

VGI metadata, i.e. data about VGI, can serve multiple purposes. As for other types of information, it can assist in the discovery, evaluation or use of VGI. While Discovery metadata elements such as title, description and keywords are commonly used directly for matchmaking with user queries, Browse or Extra elements (e.g. license, spatial accuracy, survey method) can help in judging whether a particular collection of VGI suits a particular usage scenario, and Commentary metadata such as usage citations and reviews can build trust in the product. Once a data set has been positively evaluated as of interest for a particular application, Discovery or Extra metadata which describes access protocols and storage locations can help in

[1]https://www.iso.org/standard/53798.html.

[2]https://www.iso.org/standard/66197.html.

[3]https://www.fgdc.gov/metadata/csdgm-standard.

[4]http://www.opengeospatial.org/standards/sensorml.

retrieving the relevant parts and Archive metadata allows the relevant transformations and mappings to be made in order to mobilise the data. The different use cases of VGI metadata which we present here require a rich set of different metadata elements, and it remains challenging to provide all this information about VGI. However, some components can be filled and updated automatically, and if users can access a mechanism for contributing their evaluations and opinions, this can help in completing the full set.

1.3 Existing Approaches to Crowdsourcing Metadata

Metadata crowdsourcing has been developing in a number of fields over the past decades. Volunteered archives of photographs such as Flickr Common have long allowed contributors, and sometimes others, to tag photographs with labels referring to their content, effectively providing commentary metadata about meaning. In a more systematic and commercial context, image labels can be collected from paid volunteers using platforms such as Amazon's Mechanical Turk.[5] Given enough relabelling to obtain a consistent data set, such labels can then be used as training sets for applications such as machine learning (e.g. the LabelMe initiative[6]). At this point, the tags which were associated with images and could be used to browse and discover them become effective data, identifying specific elements of an image on which a neural network can be trained. In terms of quality metadata, commercial enterprises such as Amazon, TripAdvisor and Airbnb have a longstanding history of soliciting user ratings and comments for specific resources. In the world of geospatial data, this approach has been extended to publish platforms such as Geonode[7] and ArcGIS Online,[8] which allow registered users to contribute scores and simple comments on the data layers that they make accessible. In all of these contexts, there may be no control over the contributors—often, this forces a reliance on consistency and repeated review in order to avoid the impact of malicious or mistaken individual contributions.

In the face of rapid change such as it is seen in the world of VGI and mobile data collection, it can be valuable to look to other domains for lessons on what works to engage volunteers and to ensure useful contributions. In particular, the fields of library science and cultural heritage have seen experimental moves in recent years from metadata digitized entirely by staff and 'experts' to a strategy of gathering descriptive annotations via a range of gamification strategies (e.g. [5]) such as the 'Stupid Robot' app.[9] The drivers are several: apart from the savings in terms of the

[5]https://www.mturk.com/mturk.

[6]http://labelme.csail.mit.edu.

[7]http://geonode.org/.

[8]https://www.arcgis.com.

[9]http://www.metadatagames.org/.

costs for semantic annotation by professional cataloguers, there is also huge value to be gained from opening up heritage data for inspection and discussion by the people for whom it is being curated [6]. Insights from collaborative initiatives such as the Swedish Living Archives project[10] are now being applied in the geospatial context (e.g. [7]).

When tags and annotations are completely unrestricted, they become more difficult to match to exist taxonomies of knowledge. In some cases, this has the potential to increase the richness and usefulness of a descriptive set of terms—for example, [8] found that in steve.museum—a social tagging research project for museum collections—of over 36 thousand contributed terms, 86% were completely different from the accepted vocabulary used by museum experts, but 88.2% of the contributed terms were considered by museum staff to be of use to support discovering and assessing resources online. However, without some curation and ontology development, it can become difficult to navigate and understand an increasing wealth of keywords, especially in multiple languages. For this reason, many domains restrict tags and annotations using code lists or shared vocabularies which conform to agreed standards. For example, in the biomedical domain, the CED^{2}AR project (Center for Expanded Data Annotation and Retrieval) provides tools[11] for the development and sharing of metadata templates (including sets of allowed values) between groups of scientists. Vilhuber [9] proposes a mechanism for gathering crowdsourced metadata on experiments, data sets and publications using this platform. Contributors may be authenticated and restricted (e.g. identified using an ORCID ID), and contributions might optionally be subjected to a further layer of aggregation/translation by experts.

2 Where and How Can Metadata Be Produced?

Metadata can be provided at all stages of the data lifecycle, ranging from planning and initial collection through post-processing and storage up to use, reuse and integration with other data sets. For the scope of this chapter, we primarily distinguish between metadata provision at the production stage (i.e. when VGI is actually created and provided for use; see Sect. 2.1) and the consumption stage (i.e. when VGI is actually used for the originally intended purpose or reused in another context;see Sect. 2.3). For both stages, we distinguish different metadata sources and we also discuss (in Sect. 2.2) elements of metadata which may be provided at either or both of these stages.

Independent of the stage of metadata provision, it has to be noted that it becomes increasingly important to also capture information about who provided certain pieces of information (in effect, metadata about metadata). In the emerging

[10]http://livingarchives.mah.se/.

[11]http://www2.ncrn.cornell.edu/ced2ar-web.

situations where different mechanisms and parties provide inputs to a metadata record, we have to be able to trace back information sources. Such information about the providers of (parts of) a metadata record is essential for potential data users in order to evaluate fitness for purpose. Only if such provenance information is available can the user evaluate and decide, whether they trust the information provided.

2.1 Data Production Stage

Metadata generated during data production may be *automatically generated* from the instruments or interfaces being used for collection, or *actively entered by a user*.

2.1.1 Automatically Collected

This type of metadata is usually of the Archive type, describing units, coordinates, focal length of an image, timestamps, etc., often at the level of the individual observation. It can be derived during the act of data collection by, for example a camera, GPS, sensing instrument, identity of user in Flickr, iNaturalist or Panoramio—standard fields for GeoODK). For detailed search,(e.g. where a potential user wishes to find measurements conforming to specific unit types or to filter observations by their individual location), this information also acts as Discovery metadata. This highlights an important distinction between VGI, which often consists of very granular information which can be grouped and used in a variety of combinations, and more traditional data sets which are collected and published as single entities with overarching documentation (spatio-temporal footprint, theme, etc.) to assist in discovery. There are often logical groupings to VGI such as citizen science campaigns or specific projects, but these can intersect, and very often, an observation has the potential to belong to a number of logical groupings defined by space, observer type, time window, etc.

Examples:
iNaturalist[12] is an online interface which allows volunteers to report observations of a particular species. One or more photograph must be uploaded as evidence. The user's ID is automatically attached to the record since only logged-in users may submit: but if working in the field with a mobile device, there is also the option to derive the time/date stamp and the location from that device.

The Weather Underground[13] and the UK Met Office's WOW (Weather Observation Website)[14] are two examples of networks which permit amateur

[12]http://www.inaturalist.org.

[13]https://www.wunderground.com.

[14]https://wow.metoffice.gov.uk/.

weather enthusiasts to submit data from their own weather station. A number of users upload data to both networks [10] and the observations are aggregated and interpolated to form maps and other data sets accessible to the public. Given the nature of meteorological data, the date and time are of key importance in making the data usable and are captured from the amateur weather station itself as an integral part of the observation record. This is important since connectivity limitations mean that for some users, data are cached and uploaded in packets when possible.

Key issues and challenges:
The reliability of this type of metadata may be affected by the precision and accuracy of the instrument in use—for example poor satellite coverage may distort a GPS reading or the date and time on a camera may be set incorrectly. Automated readings can also lead to error because of user error in the use of the interface: for example, if an image is not geolocated at the time of capture, but uploaded from another location, then a misleading location will be recorded. Sometimes the quality estimate from a device which forms part of the metadata is itself in error: for example, a weather station which reports a particular standard error on humidity readings but where the sensor has drifted over time because of the difficulty and expense of professional-level calibration [11]. A similar problem relating to temporal information is that which relates to the rainfall collected by a weather station, typically by a tipping bucket mechanism. Here, precipitation rate can be computed based on the number of times that the bucket fills and activates the sensor when it tips to empty itself. Thus, as well as cumulative rainfall, an estimate of rate (i.e. temporal extent of that quantity) can be achieved. However, when rainfall is low, or catch is low because of interference from wind, the bucket will not tip for a longer time and it can be harder to explicitly define the time period to which the rainfall relates [12].

Documentation of automated data transformations is exceedingly important, in order to ensure proper use and reuse of the data. Bell et al. [10] note that pressure readings are routinely corrected to sea level in the Weather Underground and WOW repositories, using the reported elevation of the station. However, some weather stations make this correction internally, before the data is published to the shared hub, and this leads to systematic bias in the final readings. A lack of provenance information from the individual sensors thus affects the quality of the aggregated data set.

Finally, it has to be recognised that automation can only result in a partial coverage of all the metadata elements that are required: to discover a data set; to evaluate if the identified data set fits an intended purpose; and to enable the use of the described data set (the three reasons why metadata is provided in the first place). For example, natural language descriptions of a data set do help a human reader to judge if a given source is of potential use, even before investigating the accuracy of spatial or thematic attributes.

2.1.2 Manual Additions by the Contributor

This type of metadata best fits the Browse and Extra categories. It is commonly captured using an analogue or digital form that elicits information from the user either through dropdowns (lists of permitted terms) or free text fields. (e.g. Darwin Core, user-generated forms for GeoODK, estimates of certainty/measurement quality (e.g. iNaturalist, WOW).

Examples:
An iNaturalist submission can be supplemented by user comments describing the context in more detail. At the point of submission, the user is free to select from a list of officially accepted species names if they are relatively confident of the identity of the species. Alternatively, if they cannot confidently identify the species, they may flag the observation as requiring help from the community for identification—an implicit assessment of current observation quality. Date and location may be entered by hand if the user is not submitting an image at the point of observation.

The Integrated Publishing Toolkit of the Global Biodiversity Information Facility (GBIF) permits a contributor to upload prepared files, (which often consist of cleaned and aggregated records from National Biodiversity Networks or the above-mentioned citizen science initiatives), and to supply metadata specifying the scope, methodology, ownership, rights, etc., of that set of records [13].

Both Weather Underground and WOW require users to submit information on the characteristics of their weather station and its placement, since shading, height above roof level and surrounding vegetation can all affect the quality of station readings.

Key issues and challenges:
The nature of metadata from different initiatives may be very different, and a lack of standardisation on elements to be recorded, and whether they should be mandatory, can make it difficult or impossible to combine data sets. Both Weather Underground (WU) and WOW require contributors to give the coordinates and elevation of their station, but elevation is occasionally missing or incorrect in both cases. The majority of users on WU supply details of their station type (a factor which is very important in predicting data quality), but this information is not requested on WOW. WOW allows users to give an estimate of the quality of their station's measurements, according to an agreed grading scheme,[15] but WU does not. Such differences mean that a valuable opportunity to fuse data from different sources across large spatio-temporal regions may be missed, without some careful data cleaning and mapping of terms [10]. By contrast, initiatives where proformas have been actively standardised offer more opportunities for data set harmonisation and meta-analysis. The Atlas of Living Australia[16] and the US Federal Crowdsourcing

[15]http://www.met.reading.ac.uk/~brugge/col/grade.html.

[16]http://www.ala.org.au/.

and Citizen Science Catalog,[17] among other initiatives, have adopted and built on the data standards developed within the SciStarter database,[18] which allows at least the projects, goals and sampling methodologies (if not yet the individual observations) to be described, discovered and compared in an easily understandable way. In terms of the individual observations, the Open Geospatial Consortium's Domain Working Group on Citizen Science is developing proposals for a profile of Sensor Web Enablement (SWE) which will be usable across a range of domains and devices.

The governance of terminologies used remains especially challenging. Whereas some purists say that only the title and description (possibly also description of how a data set has been used) should be open to free text editing, the agreement on the expressions to be used in any other metadata field can differ greatly between user groups. Restriction to code lists (e.g. for measurement units, spatial reference systems or names of observed phenomena) is often still specific to a particular user group/community. Cross-community agreements (e.g. on a registry of organisations) and registries of translations between commonly used different vocabularies would help the domain to advance beyond the current level of interoperability. Even in the absence of such agreements, third parties might still be able to provide the interconnections. An example is the thesaurus alignment implemented in support of the Global Earth Observation Systems of Systems (GEOSS) following a Linked Data approach [14].[19]

Last but not least, manual editions by definition imply human intervention. On the one hand, the required human–computer interaction is a potential source of error, which can be reduced by well-designed metadata editors (such as the INSPIRE metadata editor[20]). Nevertheless, due to the complexity of many metadata schemes, these editors also have to embrace a certain level of complexity. This might become an issue when considering many of the devices that are used for capturing VGI today, which usually feature small screens and keyboards. Even when working in a less challenging environment, the manual entry of a metadata record remains time-consuming, and contributors need the incentives for completing metadata by hand. Such incentives might include the assignment of permanent identifiers (such as Digital Object Identifiers, DOIs) that allow citation and acknowledgement; the commitment to sustain metadata for long-term use; compliance assurance with metadata standards (such as INSPIRE); and ultimately, reuse not only within the originating organisation but also by other organisations using interoperable protocols (such as CSW, CKAN API or SPARQL).

---

[17]https://crowdsourcing-toolkit.sites.usa.gov/.

[18]https://sciStarter.com/.

[19]http://www.eurogeoss-broker.eu/.

[20]http://inspire-geoportal.ec.europa.eu/editor/.

2.2 *Data Production and/or Consumption Stage*

Metadata added here is often inferred or calculated in relation to other pieces of information (e.g. from location context, relationship to neighbouring points in time or space, historical reliability of the contributor). It is possible to distinguish *inference using the same source* (e.g. using co-occurrence of observations or consistency among multiple observers) with calculations of quality or observer reputation based on correlations/overlaps/agreements with other data sets, e.g. *authoritative data sets* or *expert validations*. The flexibility of this kind of metadata means that documentation which is originally patchy can be enriched and improved throughout the life cycle of the VGI.

2.2.1 Metadata Based on Internal Consistency of the Observations

Examples:
Foody et al. [15] note the significant cost and effort required to gather independent verification data in the contexts of land cover classification and explore a latent class model method by which the internal consistency of the observers may be leveraged to more easily label high-quality observations.

Kelling et al. [16] assess the variability of their observers with reference to the rate at which they accumulate species in their recording lists. The curves representing the rate at which species are first added to a checklist and how the total levels off over time were found to be characteristic of an observer's expertise and to change as the observer becomes more experienced, especially for hard to identify species. Thus, a purely internal characteristic of the data set (combined only with some independent insights into the difficulty of identifying certain species) was successfully used to assess the likely completeness of an observer's list, as well as their general observing skill.

Key issues and challenges:
Approaches which assess consistency in this way rely on a systematic sampling design which directs contributors to comparable locations within a reasonable time frame so that change in the phenomenon being observed is avoided. Therefore, this approach is well suited to the kind of application where volunteers remotely identify or interpret imagery (e.g. the GeoWiki project) but less applicable to a citizen science programme where a user selects their own location, time and phenomenon of interest. An approach which requires the same location or entity to be sampled multiple times suffers disproportionately in locations/time frames which are rarely sampled, but in this case, an appropriate level of uncertainty may be derived from (and attached as metadata to) the original set of observations, based on its sampling pattern.

2.2.2 Metadata Based on Comparisons to External Sources

Examples:
Otegui and Guralnick [17] describe a geospatial REST API for automatically annotating GBIF records with locational accuracy and other quality flags. The process is based on a number of reference data sets such as Red List polygon species ranges, land/water masks, etc., which are considered to be adequately accurate to identify outliers and aberrations in the volunteered GBIF data.

In recent years, there has been an increasing focus on considering the assessment of the accuracy of spatial data within OpenStreetMap (OSM). Initially, several years ago, these assessments focused primarily on the geometric accuracy of OSM such as studies by [18, 19]. However, assessment of OSM accuracy has expanded to consider temporal accuracy of physical objects within the OSM database and the syntactical and semantic accuracy of tagging [20]. No clear consensus on how to compare OSM to authoritative data sets has yet emerged. This is due in no small part to the fundamental differences in how OSM is collected, why it is collected, who collects it and how the data is managed with similar procedures in authoritative data sets such as those produced by National Mapping and Cadastral Agencies [21, 18]. Each authoritative data set must be compared 'from scratch' to OSM in every case. There is unlikely to be a one-size-fits-all solution to comparison of OSM to external sources [22]. The crucial question, which has not either been sufficiently addressed or answered yet, is how the results of these comparisons to external sources can be used to assist in improving the quality of OSM data or how additional metadata can be supplied in this respect for OSM.

Spinsanti and Ostermann [23] evaluate the quality of their method to detect the use of building blocks from Foursquare by comparison with the authoritative data from a municipality, i.e. official authority. Such agreements between data sets might indicate quality but has to recognise differences in the purpose of data provision. In the given case, for example, cities are interested primarily in the use of buildings for taxation and similar purposes. Foursquare users are interested in the services provided to them (hotel, restaurant, bar, etc.). Accordingly, some degree of difference can be expected and does not determine data quality.

Murray [24] also work on data mining but concentrate on Twitter and Flickr, two other platforms that also heavily benefit from inputs from mobile devices. They deeply embed validation processes in the data production stage. Here, indications of possible forest fires are masked against areas where there is actually enough biomass (forest) to burn, and they rank outcomes using a recognised fire risk index. These checks contribute to the quality of the provided data but are deeply integrated into the detection method. Should such information appear in the metadata of the processed information or is it 'only' part of the method descriptions?

Key issues and challenges:
A key problem is the reliability of the reference data: in cases such as OSM, new data may be more informative and current than the authoritative data sets. The most recent version of OSM data for any geographical region is available in near real

time, giving it the opportunity to reflect infrastructure, urban and environmental change very quickly. However, comparison and assessment of OSM against reference data sets and vice versa must be performed in a sensitive manner due to the differences in how OSM and many reference data sets are collected and managed.

It remains essential to carefully consider and compare the purposes for which data sets (VGI ones and reference data sets) have been collected. In the case of detecting building block use from Foursquare and comparing it with data from the municipality [23], for example, Foursquare users are interested in any service that serves them directly, such as dining, living, transport, shopping. Instead, the local administration is interested in the primary, secondary, etc., use of a building block for much wider planning purposes. This is very important to note because VGI is often collected for a different purpose than existing sources and thus can complement (but not necessarily replace) them.

The semantics of the results from the comparison have to be detailed. It should be reported how exactly the data set under consideration relates to the reference data set, and how an agreement between the two can be interpreted. What does it mean if the agreement between a VGI data set and one from an authoritative source is 67%?

As already mentioned for automatically generated metadata (above), metadata derived by comparison to other data sets will not be enough to allow for data discovery, evaluation and use. It can only be a complement to other mechanisms.

2.2.3 Metadata Supplied by Expert Review

This kind of annotation or report may be supplied by a thematic expert who is not part of the data gathering community.

Examples:
Biodiversity reporting is a classic example where different communities might collect data in the field (attaching metadata including her contacts and possible usernames), and professional networks validate the collections, usually based on images and possibly including dialogues with the person who reported a specific sighting (see, e.g., Conservation International[21]). In specific cases, e.g. the detection and identification of invasive alien species, this process might have to be extended to include a final confirmation by the responsible governmental authorities since there are serious potential legal implications to the generation of incorrect records because of legal requirements for reporting and eradication. Especially with respect to validation, the classical approach of involving professional experts is increasingly complemented by approaches including a larger set of contributors. Such strategies explicitly acknowledge that the required expertise does not necessarily reside inside academia (see, e.g., [25]). It is increasingly recognised that many

[21]http://www.conservation.org.

citizens have acquired highly specialised skills because of interest in their local environment or a certain scientific niche. Accordingly, citizen science and crowd-sourcing platforms now include tools to capture related quality metadata. iNaturalist, for example, supports a relatively simple approach to validate the occurrence of a species by a simple rule (2/3 of the validators have to agree). By contrast, iSpot,[22] a website that supports anyone interested in identifying what they observe in nature, uses more sophisticated algorithms based on the history and reputation of a validating user. Supporting information has to be provided about the validation method (possibly even a chain of multiple steps) and the parties involved. This might even include a framework that allows reporting of multiple validation processes so that users can decide whether their own 'trusted validator' confirms an observation.

Key issues and challenges

Possible validation steps are manifold, and a framework would need to be flexible. A metadata structure to capture multiple validation steps is not yet in place, though within the widely used ISO 19115-3 standard for geospatial metadata; this can be achieved by adapting the 'Lineage' element to form a 'Traceability' statement which records in detail, and in a machine-readable form, all the steps of a quality assessment. This approach was proposed within the recent EU-funded project, GeoViQua.[23]

Information about organisations but also individuals needs to be processed, i.e. privacy has to be secured, sometimes by de-identification. In the wider security context, any validation involving human interaction entails the need of trust. If I, for example, want to use data from iNaturalist, how much do I trust the iNaturalist community, and the iNaturalist algorithm to provide valid information about invasive species? How much do I trust alternative communities, such as iSpot, to provide valid data? Ultimately, different users may be willing to rely on data that has been validated by communities and validation processes that they believe in and trust. Accordingly, metadata schemes might have to accommodate information about multiple validation processes for the same observation/data set.

2.3 Data Consumption Stage

This kind of metadata is provided by other community members than the collector of the data himself or herself, e.g. crowdsourced accuracy checking, annotation and feedback on resources from the end-user community who discover and use a

[22]http://www.ispotnature.org.

[23]http://www.geoviqua.org/ProducerQualityModel.htm.

resource, rather than solely from experts or volunteer validators. This type of feedback may also be on authoritative data, in which case it is the metadata which becomes the 'volunteered' resource.

Examples:
Geowiki's cropland agreement map is a distillation of feedback and annotations on original products [26]. The map does not only compile data from a set of satellite data sources but also validates against a data set that was developed using two independent crowdsourcing campaigns. Agreement analysis was done by experts coming from a research organisation as well as representatives of the Geowiki community. This is a powerful combination of classical image processing with the inputs from volunteers.

In the recent EU-funded CHARMe project [27], climate data sets were experimentally enriched with a variety of commentary metadata including publications, user feedback and reports of specific usage. Adopting the Open Annotation standard (based on Linked Data principles and utilising a Resource Description Framework (RDF) data model), the project aimed to crystallise community experience and make it accessible in such a way that conclusions can be traced back to the exact evidence which supported them. One particular finding of interest was that, while Commentary metadata is much needed and currently lacking, it should be integrated with the other, more traditional, forms of metadata in order that fitness for purpose can be properly assessed.

Key issues/challenges:
A large challenge for community-contributed metadata of this type is assessing the reliability of the commenters, identifying malicious intent and potentially excluding or weighting comments and ratings accordingly. Most proposals in the scientific or research domain involve some sort of authentication (e.g. a researcher may be identified by their ORCID) and the standard Geospatial User Feedback (OGC GUF) model and schema implementation[24] recently agreed by OGC specifies that commenters should describe their level and field of expertise, to allow filtering by domain and expertise level.

Clifford et al. [27] note that their RDF annotation approach is more flexible than the OGC GUF, but that it is consequently more difficult to query and aggregate annotations generated in this format. They propose that ultimately the two approaches may be combined, depending on the priorities and constraints of different domains. This approach is especially welcome given the existence of a third standard which addresses similar issues and which is gaining ground in the domain of non-spatial data: the W3C Dataset Usage Vocabulary.[25]

---

[24]http://www.opengeospatial.org/standards/guf.

[25]https://www.w3.org/TR/vocab-duv/.

3 Case Studies

In order to illustrate the issues outlined above, we analyse two case studies along the complete spectrum of metadata provision. The first describes the well-known OpenStreetMap as a leading example from the domain of volunteered mapping. The second outlines an application designed to identify invasive alien species across Europe, which is currently under development, as a typical case of a mobile phone app for collecting VGI. We selected these two cases because they represent highly different application areas (long-term collection by a variety of means of geographic information on many different phenomena, leading to an archive which can be subsetted in multiple ways for multiple purposes, versus a thematically focussed approach to collecting data for a very specific purpose using mobile services), and because of our close involvement with the respective developments.

3.1 Open Street Map

3.1.1 Data Production Stage: Automatically Collected Metadata

Within OSM, metadata is collected at the point of contribution for every object. There are a number of popular editor software tools which allow contributors to submit new objects to the OSM database or edit the geometry or metadata associated with existing ones. The process of contribution is easy and flexible. When a contributor creates a new object (such as a point represents a location or a polygon represents a feature) the editor software provides options for choosing appropriate tags (key–value pairs) for the description of this object [28]. The contributor is free to choose the most suitable options from this list or provide additional tags of their own. In the case of automated or 'bulk' upload of spatial data to the OSM database, there is the possibility to add data set-specific metadata automatically to each object uploaded in addition to OSM specific tags. A large number of external data sets have been inserted and uploaded to OSM in this way. OSM [29] contains an extensive listing.

3.1.2 Data Production Stage: Manual Additions by the Contribution

In many ways, the data within the OSM database is continually available for access at the data production stage. Contributors can edit the OSM database at any time. Changes are reflected almost immediately and subsequently permeate down into download services for data very quickly. When contributions are made in OSM by a single contributor a logical grouping of those contributions, called a changeset, is created in the OSM database. Contributors are strongly encouraged to provide metadata (open text) to these changes in order to document and support the changes they have made [30]. Changeset information is accessible to anyone and well-annotated changesets can provide a rich amount of contextual metadata surrounding a contributor's rationale and approach to their set of changes.

3.1.3 Data Consumption Stage: Metadata Based on Internal Consistency of the Observations

For the most part, the contributors to OSM themselves are responsible for performing their own internal consistency checks of their observations or data contributions. As outlined in [31], verifiability is an important concept in OSM and contributors are encouraged to ensure that OSM data should, as far as is reasonably possible, be verifiable. This is a good practice guideline covering all mapping activity, and also by virtue of common sense, this has become a policy governing choices about the most suitable and useful tags to use to annotate data objects. Documentation such as the Wiki pages in [32] provides a very extensive set of guides, tutorials and explanations of how contributors should approach the task of contributing new data or editing existing data in OSM. Within these documentation pages, internal consistency checks are outlined in many places. Mooney et al. [18] suggest that in the case of volunteered geographic information more standardised protocols are needed in other to encourage contributors to perform these types of internal consistency checks when delivering observations. There has been some use of 'Bots' or automated approaches to performing QA/QC checks on OSM data (such as adding tags, fixing incorrect tag keys, etc.). However, such automated approaches are not always supported and should be used with extreme caution [33].

3.1.4 Data Consumption Stage: Metadata Based on Comparisons to External Sources

The academic community has carried out many comparisons of OSM to external sources such as authoritative data sets from National Mapping and Cadastral Agencies (as outlined above in Sect. 2.2.1). Brovelli et al. [22] describe the development of a fully automated software tool which allows the quality assessment of OSM against authoritative road data sets. Quantitative measures for the completeness and spatial accuracy of OSM are computed, including the compatibility of OSM road data with other map databases. By providing users with these measures and information, it becomes the task of the user then to interpret OSM's fitness for use for applications like routing and navigation.

3.1.5 Data Consumption Stage: Metadata Supplied by Expert Review

Contributions and edits to the OSM database are available for editing and updating by any other contributor to OSM. Expert contributors to OSM could supply additional tagging or metadata to edits made by other contributors if necessary. However, given the scale of OSM and the number of edits, this is a difficult and time-consuming task. The reviews carried out by experts in spatial data management, and analysis within the academic community could be considered as possible sources of additional metadata. However, as argued above, there are no clear mechanisms by which the results of these expert reviews find their way back into the OSM database either as updated geographic data or metadata.

3.1.6 Data Consumption Stage: Community-Contributed Metadata

In the OSM ecosystem, the community contribution of metadata is very similar to the data production stage with 'manual additions by the contribution'. Dodge and Kitchin [34] comment that OSM is in a constant state of flux and can never be considered a 'final product'. This gives contributors scope to continually edit and improve both the geographical data and metadata within the OSM data.

3.2 *JRC App: Invasive Alien Species in Europe*

3.2.1 Data Production Stage: Automatically Collected Metadata

The JRC app to inform about and collect information about invasive alien species of European Union concern [35] requires user registration. It then delivers an user identifier as part of the metadata of any sighting that is submitted by the application using a secure connection. In addition, as the app offers the upload of pictures, the picture metadata is transferred, too. The exact content of the picture metadata depends to some degree on the type of phone used, but may, for example, include a second location, which might later be used as part of validation mechanisms.

3.2.2 Data Production Stage: Manual Additions by the Contributor

The data record that is sent by the app (incl. pictures, species, species density, habitat, location, date and timestamp) may also be accompanied by a comment capturing additional metadata. This is the only metadata element that can be directly edited when recording a sighting with the app.

3.2.3 Data Consumption Stage: Metadata Based on Internal Consistency of the Observations

Any metadata that is part of the quality assurance mechanisms should be recorded. Internal consistency checks still need to be developed. They will include the description and results of an automated processing of images for pre-validation, i.e. to identify species on images. The same will be provided for a language analysis of the comments in order to flag potential needs for moderation. Also, the results of consistency checks between provided locations and locations in image metadata will be included here.

3.2.4 Data Consumption Stage: Metadata Based on Comparisons to External Sources

In order to prepare the validation by human intervention, likelihoods of appearance of a species in a given habitat will be calculated. A description and the results of automated processing of species information, date and location together with known habitats will be provided as part of the metadata for quality assurance.

3.2.5 Data Consumption Stage: Inferred Metadata

Inside a protected user management system (ECAS, the official user authentication service of the European Commission), the user identifier might be used to resolve the full name of the participant and to retrieve his/her e-mail address. Assuming informed consent, this is an option if a contributor needs to be contacted as part of a validation process and may be used for this occasion or to provide feedback about the use of the inputs provided.

3.2.6 Data Consumption Stage: Metadata Supplied by Expert Review

This review of inputs is currently under development. Expert's contribution to validation including their user identifier (again using ECAS), proposed change of validation status, and comments supporting their reasoning are foreseen, together with a possible flagging for required follow-up action. We intend to provide this information in such a way that metadata about other forms of validation could also be easily added.

3.2.7 Data Consumption Stage: Community-Contributed Metadata

In the case of a mobile application, some metadata is hosted by the app stores (in this case inside Google Play and iTunes). This typically includes ratings of the app and comments. Apart from that, we currently consider involving a community in the re-validation of records, i.e. as part of the evaluation, if received pictures do indeed include species information. As in the case of an expert's validation, the corresponding metadata would include user identifier (ECAS), proposed change of validation status, comments on reasoning. We also foresee an option to validate species information with existing communities. Here, too, we would collect the ECAS user identifier, proposed change of validation status, comments on reasoning and possible flagging for required follow-up action.

4 Conclusions

In the case of OSM, we see something of a one-way street developing between OSM and other data sets such as authoritative data sets. Many studies are published indicating how OSM data is used to supplement or 'fill gaps' in other data sets for a plethora of applications. Examples include conflation of sports tracker data sets in [36, 37] for generation of Big Data gazetteers, and [38] for place name extraction and location. However, few published studies, if any, to our knowledge actually feedback updates and corrections into the OSM database. New methods of supplementing existing data objects within OSM with additional metadata are being developed—for example work by [39]. However, no clear pathway exists on how to integrate these suggested changes. The academic community must try to engage in a more meaningful way with the OSM community for such a pathway to be mutually agreed upon. Performing large-scale automated updates to OSM from these types of studies is something that needs to be properly agreed upon with the OSM community. In particular, engagement with OSM volunteer mappers active in the geographical areas that these updates are likely to affect is crucial.

Considering the case of the JRC app on invasive alien species, we witness a typical set-up where a mobile data collection app is combined with semi-automatic validation. Part of the quality assurance takes place at the earliest stage via the use of controlled vocabularies (lists) within the user interface, but likelihoods are calculated using automatic procedures after the initial data capture, and human intervention is often required as an ultimate step before data is publicly released. Any community-provided metadata is linked to third-party app stores and is not deeply embedded in the data collection and validation systems themselves. Alternatives to this rather fragmented approach are currently being discussed. These would include the involvement of the community in validation processes but also more direct feedback provision between those that collect, validate and use observation data about invasive alien species of European Union concern.

Especially when dealing with mobile devices, we identified a series of constraints for the production and consumption of metadata. This holds, for example, for the collection of manual metadata entries using small screen sizes and keyboards. The display of a wealth of metadata is equally challenging. Considering the ongoing technological transitions, we see a strong need for a dedicated elaboration and standardisation of infrastructures in order to support mobile data and metadata collection. As noted above, there are welcome steps (e.g. [27]) towards standardization in Commentary (C) metadata, and the recently initiated COST Action

'*Citizen Science to promote creativity, scientific literacy, and innovation throughout Europe*'[26] is dedicating a working group to the mapping and crosswalking of different citizen science metadata standards from around the world. Given the growing importance of multidisciplinary applications for sustainable resource use and 'Smart Cities', it is vital that we give attention to the potential future aggregation and fusion of data sets from very different domains. Interoperability of the metadata through which these data are described, discovered and evaluated will support this goal.

While this chapter focuses on the core issues for VGI metadata, we did not provide specific details on issues related to metadata management and curation. This is the topic of other ongoing work [40], where, among other issues, we discuss the importance of global unique and persistent identifiers as a central means to recognise metadata records and their individual elements and to allow for interconnections.

Acknowledgements The work on the JRC app 'Invasive Alien Species in Europe' was to a large part developed within the MYGEOSS project, which has received funding from the European Union's Horizon 2020 research and innovation programme. We thank our colleagues from the European Alien Species Information Network (EASIN) for the in-depth discussions about data validation and our colleague Lorenzino Vaccari for his inputs related to metadata interoperability.

References

1. Schade S. (2009). Semantic reference systems accounting for uncertainty. In A. Stein, W. Bijker and W. Shi (Eds.) *Quality aspects in spatial data mining*. CRC Press, Chapter 3.
2. Fonte C. C., Antoniou V., Bastin L., Estima J., Arsanjani J. J., Laso-Bayas J. -C., See L., Vatseva R. (2017, in press). Assessing VGI data quality. In G. M. Foody, L. See, S. Fritz, C. C. Fonte, P. Mooney, A. -M. Olteanu-Raimond and V. Antoniou (Eds.) *Mapping and the citizen sensor*. London, UK: Ubiquity Press.
3. Lawrence, B., Lowry, R., Miller, P., Snaith, H., & Woolf, A. (2009). Information in environmental data grids. *Philosophical Transactions of the Royal Society of London A: Mathematical, Physical and Engineering Sciences, 367,* 1003–1014. https://doi.org/10.1098/rsta.2008.0237.
4. Blower J. D., Alegre R., Bennett V. L. et al (2014) Understanding climate data through commentary metadata: The CHARMe project. In Ł. Bolikowski, V. Casarosa, P. Goodale, N. Houssos, P. Manghi, J. Schirrwagen (Eds.) *Theory and practice of digital libraries—TPDL 2013 selected workshops. Communications in computer and information science*, vol 416. Cham: Springer.
5. Ridge M. (2011). Playing with difficult objects: Game designs for crowdsourcing museum metadata. Unpublished MSc dissertation, City University, 2011. Available online at: http://www.miaridge.com/wp-content/uploads/2013/01/MiaRidge_MSc_Dissertation_OpenAccess.pdf. Accessed 19 Mar 2017.
6. Owens, T. (2013). Digital cultural heritage and the crowd. *Curator, 56,* 121–130. https://doi.org/10.1111/cura.12012.

[26]http://www.cost.eu/COST_Actions/ca/CA15212.

7. Nilsson, E. M. (2015). The smell of urban data: Urban archiving practices beyond open data. In Living Archives (Ed.), *The politics, practices and poetics of openness*. Sweden: Malmö University.
8. Trant J. (2009). Tagging, folksonomy and art museums: Early experiments and ongoing research. *Journal of Digital Information, 10*(1). ISSN 1368-7506. Available at https://journals.tdl.org/jodi/index.php/jodi/article/view/270/277. Accessed 17 Mar 2017.
9. Vilhuber L. (2016). Crowdsourcing metadata—challenges and outlook. Presentation available online at: https://ecommons.cornell.edu/bitstream/handle/1813/43887/NCRN-Spring2016-Vilhuber-Crowdsourcing-Metadata.pdf. Accessed 17 Mar 2017.
10. Bell, S., Cornford, D., & Bastin, L. (2013). The state of automated amateur weather observations. *Weather, 68,* 36–41. https://doi.org/10.1002/wea.1980.
11. Visscher, G. J. W., & Kornet, J. G. (1994). Long-term tests of capacitive humidity sensors. *Measurement Science & Technology, 5,* 1294–1302.
12. Bell, S., Cornford, D., & Bastin, L. (2015). How good are citizen weather stations? Addressing a biased opinion. *Weather, 70*(3), 75–84. https://doi.org/10.1002/wea.2316.
13. Robertson, T., Döring, M., Guralnick, R., Bloom, D., Wieczorek, J., Braak, K., et al. (2014). The GBIF integrated publishing toolkit: Facilitating the efficient publishing of biodiversity data on the internet. *PLoS ONE, 9*(8), e102623. https://doi.org/10.1371/journal.pone.0102623.
14. Fugazza C., Dupke S., & Vaccari L. (2010). Matching SKOS thesauri for spatial data infrastructures. In S. Sánchez-Alonso, I. N. Athanasiadis (Eds.) *Metadata and semantic research. MTSR 2010. Communications in computer and information science*, vol 108. Berlin, Heidelberg: Springer.
15. Foody, G. M., See, L., Fritz, S., van der Velde, M., et al. (2015). Accurate attribute mapping from volunteered geographic information: Issues of volunteer quantity and quality. *The Cartographic Journal*. https://doi.org/10.1179/1743277413Y.0000000070.
16. Kelling, S., Johnston, A., Hochachka, W. M., et al. (2015). Can observation skills of citizen scientists be estimated using species accumulation curves? *PLoS One, 10*(10), e0139600. https://doi.org/10.1371/journal.pone.0139600.
17. Otegui, J., & Guralnick, R. P. (2016). The geospatial data quality REST API for primary biodiversity data. *Bioinformatics, 32*(11), 1755–1757.
18. Mooney, P., Minghini, M., Laakso, M., Antoniou, V., Olteanu-Raimond, A. -M., & Skopeliti, A. (2016). Towards a protocol for the collection of VGI vector data. *ISPRS International Journal of Geo-Information, 5*(11), 217.
19. Girres, J.-F., & Touya, G. (2010). Quality assessment of the French OpenStreetMap dataset. *Transactions in GIS, 14*(4), 435–459.
20. Barron, C., Neis, P., & Zipf, A. (2014). A comprehensive framework for intrinsic OpenStreetMap quality analysis. *Transactions in GIS, 18,* 877–895. https://doi.org/10.1111/tgis.12073.
21. Senaratne, H., Mobasheri, A., Ali, A. L., Capineri, C., & Haklay, M. (2017). A review of volunteered geographic information quality assessment methods. *International Journal of Geographical Information Science, 31*(1), 139.
22. Brovelli, M. A., Minghini, M., Molinari, M., & Mooney, P. (2016). Towards an automated comparison of OpenStreetMap with authoritative road datasets. *Transactions in GIS*. https://doi.org/10.1111/tgis.12182.
23. Spyratos S., Stathakis D., Lutz M., Tsinaraki C. (2016). Using foursquare place data for estimating building block use. Environmental and Planning B: Urban Analytics and City Science. https://doi.org/10.1177/0265813516637607.
24. Spinsanti, L., & Ostermann, F. O. (2013). Automated geographic context analysis for volunteered information. *Applied Geography, 43,* 36–44.
25. Murray T. (2015). Tracking our butterflies and bees. Biodiversity Ireland 11.
26. Fritz, S., See, L., McCallum, I., et al. (2015). Mapping global cropland and field size. *Global Change Biology*. https://doi.org/10.1111/gcb.12838.

27. Clifford D., Blower J., Alegre R., Phipps R., Bennett V., & Kershaw P. (2014). Annotating climate data with commentary: The CHARMe project. In *Big Data from Space (BiDS'14)* (pp. 251–254).
28. Davidovic, N., Mooney, P., Stoimenov, L., & Minghini, M. (2016). Tagging in volunteered geographic information: An analysis of tagging practices for cities and urban regions in OpenStreetMap. *ISPRS International Journal of Geo-Information, 5*(12), 232.
29. OSM. (2017). A catalogue of bulk uploads to the OpenStreetMap database. http://wiki.openstreetmap.org/wiki/Import/Catalogue. Accessed 17 Mar 2017.
30. OSM. (2017). Good changeset comments in OpenStreetMap. http://wiki.openstreetmap.org/wiki/Good_changeset_comments. Accessed 17 March 2017.
31. OSM. (2017). Verifiability in OpenStreetMap. http://wiki.openstreetmap.org/wiki/Verifiability. Accessed 17 March 2017.
32. OSM. (2017). A good practice guide to contribution in OpenStreetMap. http://wiki.openstreetmap.org/wiki/Good_practice. Accessed 17 March 2017.
33. OSM. (2017). Using bots in OpenStreetMap. http://wiki.openstreetmap.org/wiki/Automated_Edits_code_of_conduct. Accessed 17 March 2017.
34. Dodge, M., & Kitchin, R. (2013). Crowdsourced cartography: Mapping experience and knowledge. *Environmental and Planning A, 45*(1), 19–36.
35. Tsiamis K., Gervasini E., D'Amico, F. et al. (2017). Citizen science application invasive alien species Europe. JRC Technical Report, JRC105285.
36. Bergman, C., & Oksanen, J. (2016). Conflation of OpenStreetMap and mobile sports trackingdata for automatic bicycle routing. *Transactions in GIS, 20,* 848–868. https://doi.org/10.1111/tgis.12192.
37. Gao, S., Li, L., Li, W., Janowicz, K., & Zhang, Y. (2017). Constructing gazetteers from volunteered big geo-data based on hadoop. *Computers, Environment and Urban Systems, 61,* 172–186. https://doi.org/10.1016/j.compenvurbsys.2014.02.004.
38. Vasardani, M., Winter, S., & Richter, K. -F. (2013). Locating place names from place descriptions. *International Journal of Geographical Information Science, 27,* 2509–2532.
39. Basiri, A., Amirian, P., & Mooney, P. (2016). Using crowdsourced trajectories for automated OSM data entry approach. *Sensors, 16*(9), 1510.
40. Bastin, L., Schade, S., & Schill, C. (2017). Data and metadata management for better VGI reusability. In G. M. Foody, L. See, S. Fritz, C. C. Fonte, P. Mooney, A. -M. Olteanu-Raimond, & V. Antoniou (Eds.), *Mapping and the citizen sensor*. London, UK: Ubiquity Press. https://doi.org/10.5334/bbf.k.

Data Quality Assessment in Volunteered Geographic Decision Support

Guy De Tré, Robin De Mol, Sytze van Heteren, Jan Stafleu, Vasileios Chademenos, Tine Missiaen, Lars Kint, Nathan Terseleer and Vera Van Lancker

Abstract Geographic decision support systems aim to integrate and process data originating from different sources and different data providers in order to create suitability models. A suitability model denotes how suitable geographic locations are for a specific purpose on which decision-makers need to make a decision. Particularly in the presence of volunteered information, data quality assessment becomes an important aspect of a decision-making process. Geographic data are commonly prone to incompleteness, imprecision and uncertainty, and this is even more the case with volunteered data. To correctly inform the users, it is essential to communicate not only the suitability degrees highlighted in a suitability model, but also the confidence about these suitability degrees as can be derived from data quality assessment. In this chapter, a novel hierarchical approach for data quality assessment, supporting the computation of associated confidence degrees, is introduced. To illustrate its added value, aspects of the project Transnational and Integrated Long-term marine Exploitation Strategies (TILES) are used. Providing confidence information adds an extra dimension to the decision-making process and leads to more sound decisions.

G. De Tré (✉) · R. De Mol
Department of Telecommunications and Information Processing, Ghent University, Sint-Pietersnieuwstraat 41, B9000 Ghent, Belgium
e-mail: guy.detre@ugent.be

S. van Heteren · J. Stafleu
TNO—Geological Survey of the Netherlands, Princetonlaan 6, 3584 CB Utrecht, The Netherlands

V. Chademenos · T. Missiaen
Department of Geology, Ghent University, Krijgslaan 281 S8, B9000 Ghent, Belgium

L. Kint · N. Terseleer · V. Van Lancker
Royal Belgian Institute of Natural Sciences, Natural Environment, Gulledelle 100, B1200 Brussels, Belgium

G. Bordogna and P. Carrara (eds.), *Mobile Information Systems Leveraging Volunteered Geographic Information for Earth Observation*, Earth Systems Data and Models 4, https://doi.org/10.1007/978-3-319-70878-2_9

1 Introduction

Decision support systems (DSS) aim to facilitate decision-makers in performing difficult decision tasks. A commonly used approach in decision support relies on multi-criteria decision methods (MCDM) [1, 2]. Underlying MCDM is the assumption that decision-makers first specify multiple criteria to express their preferences regarding the different options that are compared to the decision-making. Second, the criteria are evaluated for each option. And third, evaluation results are compared to each other to select the best option regarding the decision-maker's preferences.

MCDM has been successfully applied in geographical decision-making with decision-makers having to decide about the suitability of a set of given geographic locations (or geographic areas) for a specified purpose (e.g., land-use, urban development, agriculture). Commonly used comparison methods of MCDM in location-based decision support are based on simple additive scoring (SAS, also known as SAW, simple additive weighting) [3], ordered weighted average (OWA) [4], outranking methods (ELECTRE, PROMETHEE) [5], the multi-attribute value technique (MAVT) [6], the multi-attribute utility technique (MAUT) [7], the analytic hierarchy process (AHP) [8], or logic scoring of preference (LSP) [9]. A comparative study on these methods can be found in [10] and [11]. In general, each of these methods computes an overall suitability degree for each geographic location under consideration, taking as inputs the criterion evaluation results with the data of that location. 'Competing' locations can then be compared on the basis of their overall suitability degree.

A problem with geographical MCDM is that properly evaluating all criteria for all options requires a huge amount of data, which might not all be available, nor be of the same quality. This is especially the case if (parts of) the data originate from different organizations that offer their data on a volunteered basis. Commonly, not all volunteered data serve the same purposes, rely on the same techniques, or have the same precision for every location/area under consideration. As an initial example, consider sampling to establish the substrate composition of the seabed for the purpose of sustainable resource management in the project Transnational and Integrated Long-term Marine Exploitation Strategies (TILES) [12] that will be further described in this chapter. Different areas in the sea are sampled with a different sampling density. For example, there are more sediment samples in and around a harbour than in an area far away from the coast. Seabed samples are taken by various parties for different purposes. For example, construction companies collect samples for stability studies, the government collects samples for the purpose of environmental monitoring, while extracting companies might collect samples for resource-quality assessment. Each party can voluntarily share (part of) its data with the others. The varying confidence in sample quality and analytical soundness propagates to a varying confidence in computed suitability degrees. Figure 1 illustrates the influence of sample density. Samples in and around the harbour of Zeebrugge are shown. The farther we move away from the harbour, the

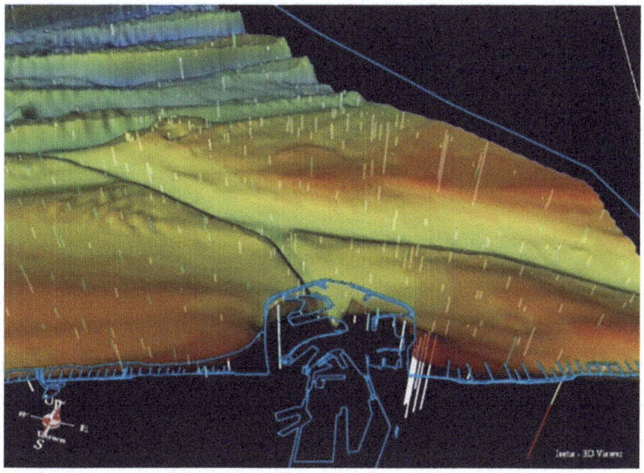

Fig. 1 Sample distributions in the vicinity of the harbour of Zeebrugge

fewer samples are available. Sampling data originate from different parties and different data sources [13, 14].

In this chapter, we describe how confidence in computed suitability degrees can be estimated and properly handled. The proposed solution consists of a novel technique that assesses data quality and provides an additional confidence degree for each computed suitability degree. In this way, decision-makers are provided with additional information needed to end up with optimal solutions. Because data quality assessment is a complex issue involving multiple aspects, data quality must be considered from a multi-dimensional perspective. The data quality of each data item used for criteria evaluation is characterized by a number of elementary components. For example, elementary quality aspects of sample descriptions include the uncertainty about the lithological classification used to characterize the sample, the sampling method and the sampling date [15]. These and other quality aspects are the inputs for the computation of an aggregated quality indicator. For example, the confidence in the lithology (gravel, sand or clay/mud) assigned to a given location depends on the availability of soil samples in the neighbourhood and can be estimated from the elementary quality aspects of these samples and their associated descriptions and analyses. In a similar way, the overall confidence in a computed suitability degree can be estimated from the quality aspects of all data required for the computation of the suitability degree. All data quality aspects are configured in accordance with the decision-maker's preferences and include an indication of their relative importance to (or impact on) the overall confidence measure. In this way, the subjective character of data quality can be taken into account.

The remainder of this chapter is organized as follows: Section 2 addresses the TILES project, Sect. 3 gives some preliminaries on the creation of suitability maps, and Sect. 4 deals with the specification of elementary aspects of data quality assessment, the evaluation of elementary quality aspects and the aggregation of quality aspects. In Sect. 5, we show how the concept of a suitability model can be enriched with confidence measures. Following some practical experiences drawn from the TILES project, in Sect. 6, we provide conclusions of this work, some open challenges and directions for further research in Sect. 7.

2 The TILES Project

The project Transnational and Integrated Long-term marine Exploitation Strategies (TILES) is funded by the Belgian federal government. It is a multidisciplinary cooperative effort between the Royal Belgian Institute of Natural Sciences (Directorate Natural Environment), TNO—Geological Survey of the Netherlands, and Ghent University (Renard Centre of Marine Geology and Database, Document and Content Management research group) [12].

One of the aims of the TILES project is to collate data and metadata and to develop IT tools that support decision-making related to the sustainable extraction of offshore sand and gravel from the Belgian and southern Netherlands part of the North Sea subsurface. These geological resources can be considered to be non-renewable on timescales relevant for decision-makers. The sustainable management of these resources requires a thorough and careful balancing of available quantity and quality versus rapidly changing societal and economical needs. The need for such an approach is recognized in the European Union's Raw Materials Initiative, which highlights the optimization of a geological knowledge base as a key element in ensuring sustainable supplies from within the EU borders. Comprehensive knowledge of the distribution, composition and dynamics of geological resources therefore is the backbone of long-term strategies for resource use in a rapidly changing world [12].

2.1 Decision Support Tool

Geographic MCDM relies on the availability of several IT components, including databases and other data resources, a data integration component, a decision support tool and a visualization tool (Fig. 2). Data from different sources are used for evaluating the decision criteria. In a first step, each data source is subject to a kind of Extract, Transform and Load (ETL) process, which extracts the relevant data from the source, transforms it into the correct format and prepares it as input for the decision support tool. For main data sources that do not change often, like 3D lithological classification models, ETL is done in advance. In the next step, the integrated input data are processed by the decision support tool. In TILES, an LSP-based approach [9, 16] is used for the computation of suitability degrees for a 3D suitability model. This model is enriched with metadata expressing the confidence that experts have in the computed suitability degrees. The enriched suitability model is loaded into dedicated visualization software for easily accessible presentation to the decision-maker.

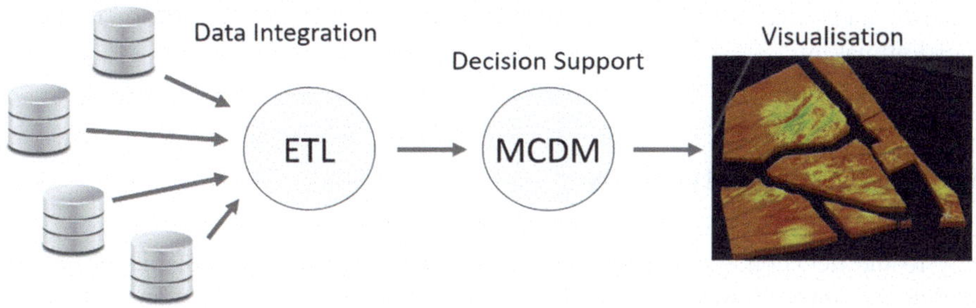

Fig. 2 Most important IT components for decision support in TILES

2.2 Data Sources

Geographical decision-making is often complicated by data imperfections. Aside from missing or incomplete data, available data might be imprecise and/or uncertain. Discarding imperfect data is commonly not an option because precise and certain data are in short supply and because imprecision and uncertainty are subjective (depending on the context/user a piece of information might be considered as being precise or not). Offshore seabed sampling is in any case time-consuming and expensive.

Uncertainty is, among others, caused by the fact that there are multiple data sources hosted by different data providers, some of them voluntarily offering their data. In TILES, various types of sample descriptions and analyses (providing information on the seabed composition) [13, 14], bathymetrical data (providing information on water depth, e.g., from Flanders Hydrography, http://www.vlaamsehydrografie.be/), and location maps of exclusion zones in and around shipwrecks, supply lines, military areas or areas dedicated to power production (e.g., from Flanders Hydrography and Belgian Marine Atlas, https://odnature.naturalsciences.be/marine-atlas/), need to be merged.

The lithological descriptions are particularly prone to imprecision and uncertainty [15]. Not all samples are taken by the same data provider, some samples are old and the seabed surface changed constantly, different sampling methods were used, different sample equipment was used, weather conditions during sampling varied, the quality of the samples themselves varied, and, importantly, different people describe samples differently. Even for data originating from the same data provider, the quality varies in predictable and unpredictable ways.

For these reasons and because we preferably should not discard any data, it is important to provide extra information to the user, indicating to what extent different parts of a produced suitability map can be trusted (or not). Else, the computations cannot be correctly interpreted and the decision-maker might be misled. This shortcoming is our main motivation for data quality assessment and for the proposition of a novel confidence measure in TILES. The proposed techniques are generic and relevant for any application of geographic information in decision-making.

3 Basic Suitability Maps

Key to the decision support tool used in TILES is the creation of 3D suitability models. As input, a 3D voxel model of the subsurface of the Belgian part of the North Sea is being constructed. Each voxel is currently fixed in size (length and width 200 m, height 1 m). The model currently includes voxels that range from 4 m above mean sea level to 75 m below it. In our previous research, we studied 2D suitability maps constructed using the LSP approach [16].

In this work, we used the same technique to create 3D suitability models, as depicted in Fig. 3. It entails a three-step method consisting of (1) elementary attribute selection, (2) elementary criteria specification and evaluation and (3) aggregation.

3.1 Elementary Attribute Selection

The construction of suitability models starts with identifying the data on which the decision process will be based. To facilitate this process, a hierarchical top-down approach is followed. General aspects are further detailed in more specific ones, resulting in a so-called attribute tree. The simplified attribute tree shown in Fig. 4 emphasizes that for a marine exploitation decision (root node), one has to consider data that are related to the lithology, bathymetry, ecology, restricted areas and locations of shipwrecks. As a measure of lithology, the probabilities for different kinds of sediment (in the example limited to fine sand, coarse sand and clay) are

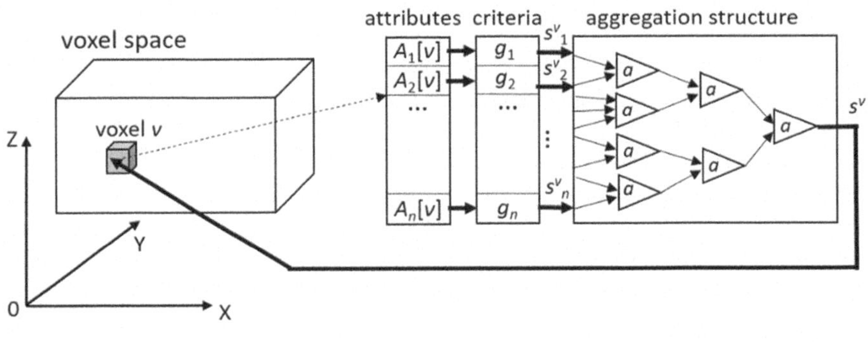

Fig. 3 LSP suitability model construction

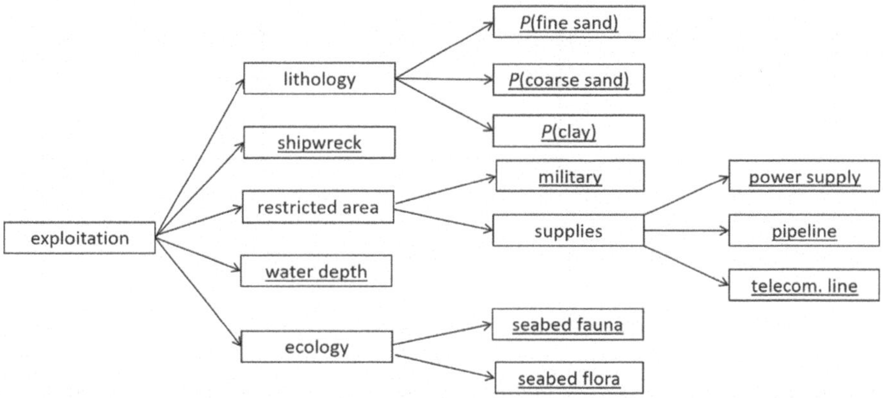

Fig. 4 Attribute tree for marine exploitation

considered. A restricted area can be a military zone or an area where power supplies, pipelines or telecommunication lines are located.

The leaf nodes of an attribute tree are the elementary attributes of the decision support process. Adequate sources must be available for assigning data to these attributes, for every 3D voxel under investigation. The other nodes represent composed attributes.

Considering a voxel v in a 3D (geological) model space, the elementary attributes can be denoted by $[A_1,\ldots,A_n]$, $n \in \mathbb{N}$ as depicted in Fig. 3 and their respective values for a voxel v will be written as $A_i[v]$, $i = 1,\ldots,n$.

3.2 Specification and Evaluation of Elementary Criteria

For each elementary attribute A_i, $i = 1,\ldots,n$, an elementary decision criterion $g_i : \text{dom}_{A_i} \rightarrow [0, 1]$ is specified. This criterion expresses the preferences of the decision-maker with respect to the value of the attribute and is defined by a fuzzy set [17] over the domain dom_{A_i}, i.e., the set of allowed values for A_i. For a given value $x \in \text{dom}_{A_i}$, $g_i(x)$ then expresses to which extent ($\in [0, 1]$), according to the decision-maker, x is a preferred value for A_i (a 0 denotes not preferred, whereas 1 reflects fully preferred).

For a given voxel v, each criterion g_i is evaluated using the actual value $A_i[v]$ of A_i for v, resulting in a so-called elementary suitability degree $s_i^v = g_i(A_i[v])$. In Fig. 5, the elementary criteria for the elementary attributes '*P(fine sand)*', '*P(coarse sand)*' and '*P(clay)*' are specified. The criterion for '*prob. fine sand*' expresses that the decision-maker prefers voxels with more than 80% probability of fine sand, although between 70 and 80% is still acceptable. The decision-maker's preference for '*P(coarse sand)*' is less than 10%, but with 10–20% still acceptable.

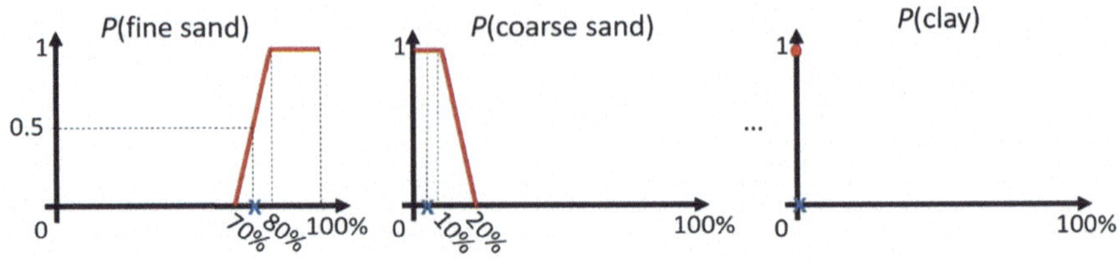

Fig. 5 Some elementary criteria and their evaluation

For '*P*(*clay*)', the preference is strict: only voxels with no clay, i.e., a probability of 0%, are allowed. For a given voxel v, criterion evaluation is done by computing the membership grade of the actual value of the attribute under consideration. Assuming actual measured attribute values $P(\text{fine sand})[v] = 75\%$, $P(\text{coarse sand})[v] = 6\%$ and $P(\text{clay})[v] = 0\%$, as depicted by 'x' symbols in Fig. 5, the elementary suitability degrees $s^v_{P(\text{fine sand})} = 0.5$, $s^v_{P(\text{coarse sand})} = 1$ and $s^v_{P(\text{clay})} = 1$ are obtained.

3.3 Aggregation

In the final step, all elementary suitability degrees s^v_i are aggregated to an overall suitability degree s^v for voxel v. For this purpose, a preconfigured, so-called LSP aggregation structure is used [16]. The LSP aggregation structure is configured in such a way that it reflects (and implements) the attribute tree structure that is constructed during elementary attribute selection. It is composed of LSP aggregators that take a number of suitability degrees as input to compute an aggregated suitability degree. Both basic and advanced aggregators can be used.

A basic aggregator $a^r_{w_1,\ldots,w_n} : [0, 1]^n \to [0, 1]$ n elementary suitability degrees s^v_i, $i = 1, \ldots n$, as input and is preconfigured with n weights $w_i \in [0, 1]$, $i = 1, \ldots, n$ and a parameter $r \in [-\infty, +\infty]$, which determines the logical behaviour of the aggregator. The weight w_i corresponds to the elementary suitability degree s^v_i and denotes the relative importance of this input (i.e., compared to the other inputs) in the aggregation. Moreover, $\sum_{i=1}^{n} w_i = 1$. If $r = -\infty$, then the aggregator behaves like a full conjunction ('and'), whereas for $r = +\infty$ a full disjunction ('or') is obtained. With the intermediate parameter values, a continuous transition from conjunction to disjunction is modelled. This is why $a^r_{w_1,\ldots,w_n}$ is called a generalized conjunction/disjunction (GCD) aggregator. A commonly used implementation of GCD is based on weighted power means (WPM):

$$a^r_{w_1,\ldots,w_n}(s_1,\ldots,s_n) = \begin{cases} \left(w_1 s_1^r + \ldots + w_n s_n^r\right)^{1/r}, & \text{if } 0 < |r| < +\infty \\ \min(s_1,\ldots,s_n), & \text{if } r = -\infty \\ \max(s_1,\ldots,s_n), & \text{if } r = +\infty \end{cases} \quad (1)$$

Advanced aggregators are provided for combining a mandatory with a non-mandatory input. A conjunctive partial absorption CPA aggregator combines a mandatory with a desired input, whereas a disjunctive partial absorption DPA aggregator is used to combine a sufficient input with a desired input. Herewith, satisfaction (resp. dissatisfaction) of the desired input results in a reward (resp. penalty) being considered in the computation of the resulting suitability degree [18].

Once the overall suitability degree of a voxel is computed using the LSP aggregation structure, it is visualized. Visualization can be done using a coloured gradual scale where red represents not suitable at all, green denotes completely suitable and intermediate colours denote partial suitability. By doing so for every voxel in the 3D area, a coloured 3D suitability model is produced. This model indicates how suitable different locations are, considering the decision-maker's criteria and preferences. Suitability maps can be dynamically generated at runtime (using different parameters and criteria), thus supporting decision-makers while checking different scenarios and reasoning processes rapidly and interactively.

4 Data Quality Assessment

The construction of basic suitability maps, as described in the previous section, works well under the preconditions that all data are available and additionally are of a satisfactory quality. These preconditions are usually not met in practice. In TILES, for example, lithological descriptions are only available for borehole locations. For other locations, the data have to be estimated by interpolation and subsequent interpretation by geologists, who apply their domain knowledge for manually correcting the interpolation results if needed. Moreover, borehole data are provided by different volunteering providers, who collect samples for different purposes at different locations, at different times, and using different methods. Finally, sample description introduces an inherent element of subjectivity, even when standard methods are followed. Hence, the data quality varies strongly among the locations under consideration.

Variability in data quality will propagate in the evaluations of the criteria and in the aggregations of the suitability degrees. So, the resulting (overall) suitability degrees cannot reflect the actual suitability of each voxel with equal accuracy. For locations with poor data quality, it is highly likely that the computed suitability degree is an over- or underestimation. If ignored, this inaccuracy can have a drastic impact on decision-making and lead to wrong decisions. To avoid misleading the decision-maker by suggesting certainty, extra information on the trustworthiness of the computed suitability degree should be provided for each voxel. This trustworthiness can be expressed by a confidence degree, which can be a value of the unit interval $[0, 1]$ and should meaningfully reflect the impact of data quality on the computations of the suitability degree.

Hence, confidence degrees should help to improve the correctness of decision-making. A high confidence indicates that the computed suitability degree is likely an accurate reflection of the real suitability of the location, whereas a low confidence denotes that the calculated suitability degree might differ severely from reality because of poor data quality. Decision-related conclusions cannot be safely drawn from such suitability degrees and the data under consideration should be improved or supplemented. In TILES, for example, an area with high suitability but low confidence due to a lack of high-quality lithological data might be seen as having high potential, but needing further investigation through extra boreholes or otherwise.

To incorporate this data-related uncertainty, we propose a novel technique to assess the quality of 'lithology' data, along with the same line used to create basic suitability models. Key to this process is finding out how (criteria for) elementary data quality aspects can be specified, evaluated and aggregated to an overall confidence degree.

4.1 Elementary Data Quality Assessment Aspects

In order to assess data quality, the attribute tree that is used for creating suitability models is considered. For each elementary attribute in the tree, a decision-maker can specify one or more data quality aspects. A data quality aspect denotes a characteristic of (the data source for) the attribute thought to influence the quality of the attribute's (computed) value. Data quality aspects can be hierarchically structured in order to model their generalization/specialization characteristics. In this way, the multi-dimensional and subjective nature of data quality assessment is reflected. Leaf nodes in such a tree structure represent elementary data quality aspects, whereas internal nodes denote composed aspects.

For example, reconsider the attribute tree in Fig. 4. The decision-maker can decide that with the elementary attributes '*P(fine sand)*', '*P(coarse sand)*' and '*P (clay)*', the same data quality aspects '*sampling conditions*' and '*model uncertainty*' have to be considered, where '*sampling conditions*' is further specified by the '*sampling time*' and the '*sampling method*'. For the elementary attribute '*shipwreck*', the '*positioning imprecision*' (in metres) is relevant. Data on the composed attribute '*restricted area*' are assumed to be adequately known. For the elementary attribute '*water depth*', the '*model uncertainty*' and '*measurement time*' are important, whereas for '*seabed fauna*' and '*seabed flora*' data the same data quality aspect '*uncertainty score*' (provided by marine ecologists) is considered.

Data quality aspects for attributes can be depicted in an enriched attribute tree as shown in Fig. 6. Each aspect is depicted by an oval in dotted lines and connected

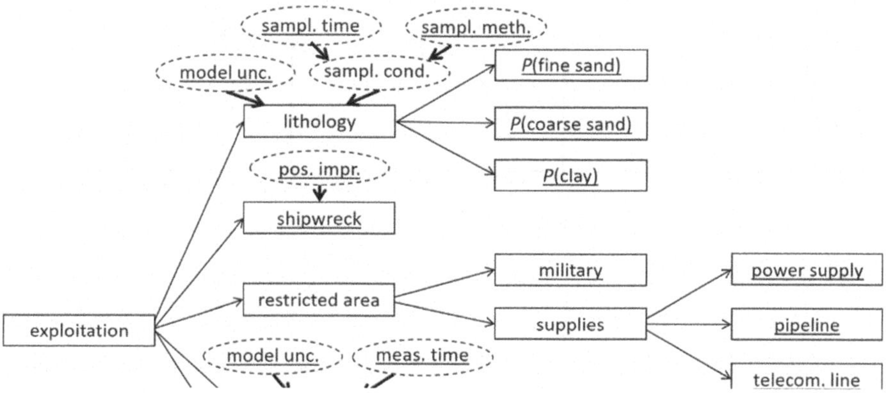

Fig. 6 Enriched attribute tree for marine exploitation

via a solid line with its associated elementary attribute. Elementary data quality aspects are underlined. To avoid overloading the tree, in case the same data quality aspect applies to all elementary attributes of a composed attribute, its corresponding oval might be connected to the composed attribute instead. Examples are the ovals connected to '*lithology*' and '*ecology*'.

For each elementary data quality aspect, relevant metadata must be provided. These metadata can either be derived from the data, or be explicitly provided by the data provider. For example, model uncertainty can be estimated from the process used to generate the lithological model, whereas metadata like sampling time and sampling method are usually given with the data source describing the samples. Considering a voxel v in a 3D (geological) voxel space, the elementary data quality aspects can be denoted by $[Q_1, \ldots, Q_m]$, $m \in \mathbb{N}$ and their respective values for voxel v will be written as $Q_i[v]$, $i = 1, \ldots, m$.

4.2 *Evaluating Criteria for Elementary Data Quality Aspects*

For each elementary data quality aspect Q_i, $i = 1, \ldots, m$, a corresponding elementary quality criterion $g_i^Q : \mathrm{dom}_{Q_i} \rightarrow [0, 1]$ has to be specified. This criterion expresses the decision-maker's acceptance and rejection preferences with respect to Q_i and is defined by a fuzzy set over the set of allowed values for Q_i (i.e., dom_{Q_i}). For a given value $x \in \mathrm{dom}_{Q_i}$, $g_i^Q(x)$ then expresses to which extent ($\in [0, 1]$) the decision-maker accepts x as being a confident value for the data quality aspect Q_i. A 0 denotes not confident, whereas 1 reflects fully confident.

For a given voxel v, each elementary criterion g_i^Q is evaluated using the actual value $Q_i[v]$ of Q_i for v, resulting in a so-called elementary confidence score

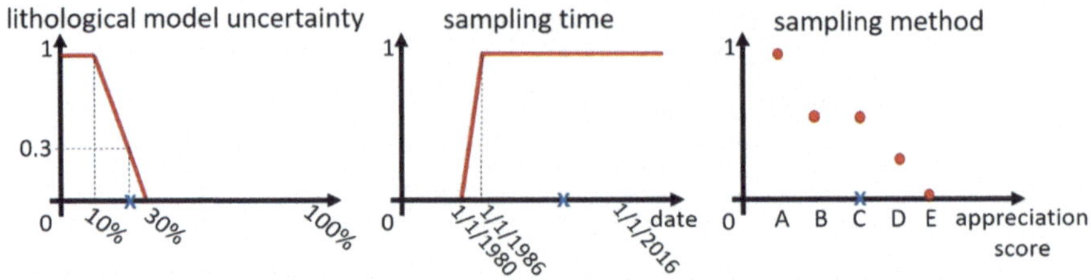

Fig. 7 Some criteria for elementary quality aspects and their evaluations

$c_i^v = g_i^Q(Q_i[v])$. This score reflects to which extent the decision-maker has confidence in the quality of the data that are related to voxel v (and are used to compute the suitability of v) as far as the elementary data quality aspect Q_i is considered. Three examples of criteria for elementary quality aspects and their evaluations are given in Fig. 7.

The left membership function models the criterion for lithological model uncertainty, which is estimated during geological 3D subsurface modelling for each voxel under consideration and is expressed by a probability denoting the uncertainty about the lithological classification for that voxel. With this criterion, the decision-maker expresses full confidence in voxels with a model uncertainty that is lower than 10%, partial confidence in model uncertainties between 10 and 30% and no confidence in model uncertainties larger than 30%. The two other membership functions model data quality criteria that are related to metadata of subsurface samples, more specifically, the oldest sampling date of neighbouring samples and the overall precomputed appreciation score for the sampling methods used to collect neighbouring samples (five possible scores 'A', 'B', 'C', 'D' and 'E' are considered). Actual values for a given voxel v for the three quality aspects are presented with a cross. Evaluating the criteria with these values, respectively, results in an elementary confidence score 0.3, 1 and 0.5.

An important challenge here lies in developing a representative technique to assess each elementary data quality aspect under consideration. An example of such a technique, used in TILES, is detailed in Sect. 6.

4.3 Aggregated Data Quality Aspects

Quite similar to how the global suitability of a system is calculated by aggregating elementary suitability degrees, a global degree of confidence is computed by aggregating elementary confidence degrees. It is done for each voxel under consideration and performed in two steps. Consider a voxel v. In the first step, a single confidence score is computed for each elementary attribute using the data and metadata related to v. In the second step, all confidence scores for v become extra inputs of the aggregation structure for the computation of overall suitability degrees, in which the aggregators are adapted in order to properly compute the overall confidence degree for v.

4.3.1 Computing a Confidence Degree Per Elementary Attribute

Considering a voxel v of a 3D voxel space, a single confidence degree is determined for each elementary attribute of the (enriched) attribute tree. The following cases can occur.

- **One elementary data quality aspect Q is specified for the attribute.** Such a data quality aspect is either directly attached to the attribute in an enriched attribute tree (like '*positioning imprecision*' is attached to '*shipwreck*' in Fig. 6) or is indirectly attached via a composed attribute of which the attribute is an element (like '*uncertainty score*' is indirectly attached to '*seabed fauna*' in Fig. 6). In this case, the elementary criterion g^Q for the aspect Q is evaluated, and the resulting value $c^v = g^Q(Q[v])$ becomes the elementary confidence score for the attribute.
- **Multiple elementary data quality aspects Q_i, $i = 1, \ldots, k$ are specified for the attribute.** Quality aspects can be directly or indirectly attached to the attribute in an enriched attribute tree and some of them may be part of a composed data quality aspect. In this case, each corresponding elementary criterion g_i^Q, $i = 1, \ldots, k$ is evaluated and the resulting confidence values $c_i^v = g_i^Q(Q_i[v])$ have to be aggregated into a single confidence score c^v for the attribute. For the aggregation, a hierarchic aggregation structure with basic GCD aggregators $a^r_{w_1,\ldots,w_l} : [0, 1]^l \rightarrow [0, 1]$, as defined by Eq. (1), can be used. The decision-maker has to configure this aggregation structure by providing appropriate weights w_i, $i = 1, \ldots, l$ and an appropriate logical behaviour parameter r for every aggregator in the structure. As an example, consider the computation of the overall confidence score for the elementary attribute '*P(fine sand)*' in Fig. 6. This attribute has two data quality aspects: the elementary quality aspect '*lithological model uncertainty*' and the composed quality aspect '*sampling condition*', which consists of the two elementary quality aspects '*sampling time*' and '*sampling method*'. The three elementary quality aspects can be evaluated using quality criteria as the ones presented in Fig. 7. Next, the resulting confidence scores $c^v_{\text{sampl. time}}$, $c^v_{\text{sampl. method}}$ and $c^v_{\text{model unc.}}$ are aggregated using an aggregation structure as depicted in Fig. 8. This aggregation structure reflects the composed nature of '*sampling condition*' for which the decision-maker considers both components '*sampling time*' and '*sampling method*' equally important. It also reflects that '*sampling condition*' is considered to be less important than '*lithological model uncertainty*'. Both aggregators are chosen to be full conjunctions (i.e., $r = -\infty$).

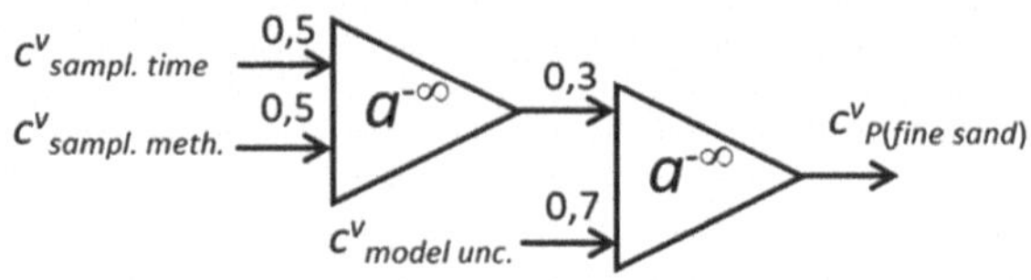

Fig. 8 Aggregation tree for the elementary data quality aspects of the elementary attribute '*P*(*fine sand*)'

- **No elementary data quality aspects are specified for the attribute.** In this case, the decision-maker has to assign an ad hoc confidence degree to the attribute. In case the data are considered to be adequate enough an ad hoc confidence degree of 1 can be assigned. As such, with respect to the elementary attributes of the composed attribute '*restricted area*' in Fig. 6, $c^v_{\text{power supply}} = c^v_{\text{pipeline}} = c^v_{\text{telecom. line}} = c^v_{\text{military}} = 1$. Other choices for confidence degrees are possible.

4.3.2 Computing the Overall Confidence Degree

The confidence degrees $c^v_i, i = 1, \ldots, n$ for the elementary attributes $A_i, i = 1, \ldots, n$ are the extra inputs for the aggregation structure that is used for computing the overall suitability s^v of voxel v. However, in order to process these inputs and additionally compute the overall confidence degree c^v for voxel v, the aggregators have been extended. Our novel, extended process for creating suitability models is schematically represented in Fig. 9.

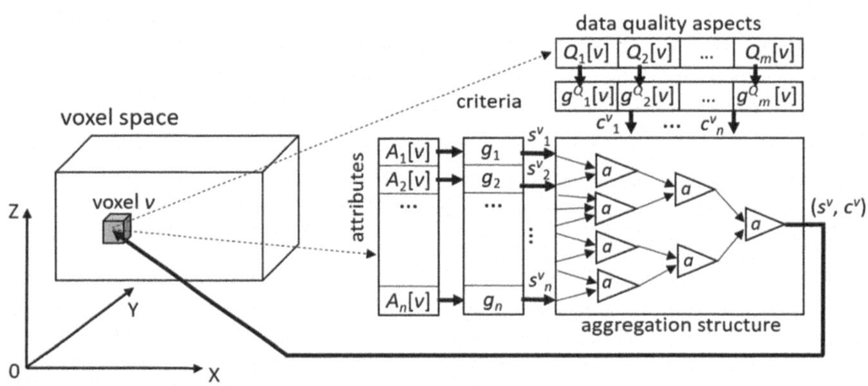

Fig. 9 Method to create an extended suitability model

The definition of the extended aggregators is an extension of the definition of the GCD aggregators as described in Sect. 3.3 (cf. Equation (1)) and is given as follows:

$$ga^r_{w_1,\ldots,w_n}(s_1,\ldots,s_n;c_1,\ldots,c_n) = (s,c) \tag{2}$$

where

$$s = \begin{cases} (w_1 s_1^r + \ldots + w_n s_n^r)^{\frac{1}{r}}, & \text{if } 0<|r|<+\infty \\ min(s_1,\ldots,s_n), & \text{if } r=-\infty \\ max(s_1,\ldots,s_n), & \text{if } r=+\infty \end{cases}$$

and

$$c = \frac{w'_1}{\sum_{i=1}^n w'_i} c_1 + \ldots + \frac{w'_n}{\sum_{i=1}^n w'_i} c_n$$

with $w'_i = d_{\max} - |s - s_i|$ and $d_{\max} = \max(s_1,\ldots,s_n) - \min(s_1,\ldots,s_n)$.

The motivation for using Eq. (2) is the following. Each aggregator $ga^r_{w_1,\ldots,w_n}$ has $2n$ inputs and generates 2 outputs: n inputs s_i, $i = 1,\ldots,n$ are suitability degrees and are used to compute an aggregated suitability degree s, the n other inputs c_i, $i = 1,\ldots,n$ are confidence degrees and are used to compute an aggregated confidence degree c.

The aggregation of the suitability degrees is done by using a GCD aggregator in exactly the same way as in Eq. (1). So, the considered sdata quality aspects have no impact on the computation of suitability degrees.

For the aggregation of the confidence degrees, a different strategy is proposed. The core consideration is that the impact of a suitability degree s_i on the computation of the overall suitability degree s determines the impact of its associated confidence degree c_i on the computation of the overall confidence degree c. Because of the use of a GCD aggregator $ga^r_{w_1,\ldots,w_n}$, the impact of a suitability degree s_i on the aggregation result s is determined by its associated weight w_i and the parameter r, which determines the logical behaviour of the aggregator. It can be estimated by the difference $|s - s_i|$. The smaller this difference is, the larger the impact will be. Hence, a weight $w'_i = d_{\max} - |s - s_i|$ can be considered for each c_i, where $d_{\max}$ is the largest possible difference between an input and the outcome of the aggregator. The property of internality, which states that the output of an aggregator is bound by the minimum and maximum of its inputs, holds for a GCD aggregator [18]. Hence, $d_{\max} = \max(s_1,\ldots,s_n) - \min(s_1,\ldots,s_n)$. Finally, to model the impact of the weights w'_i, a weighted average is proposed for the computation of the aggregated confidence degree c. By using a neutral weighted average aggregator, we opted to avoid large semantic discrepancies. But the logical behaviour of the GCD aggregator for the suitability degrees is not reflected by this weighted average. This is an interesting topic for further research of which some initial results have already been presented in [19].

5 Enriched Suitability Models

The methodology described above can be applied to compute a (s^v, c^v) pair for each voxel v that is involved in a geographical decision-making process. The value s^v represents the overall suitability of that voxel for the decision purpose under consideration, whereas the value c^v reflects to which extent the suitability degree s^v can be considered as accurate taking into account the specified data quality aspects.

A specific challenge is the graphical representation of the (s^v, c^v) pairs. Obvious options are:

- **Two separate maps.** In this option, a suitability map and a confidence map are constructed. The voxels in the suitability map are appropriately coloured in accordance with their computed overall suitability s^v, whereas the voxels in the confidence map are coloured on the basis of their overall confidence c^v. Both maps are offered to the end user who can use them for decision-making.
- **One map using a global quality measure.** The overall suitability s^v and confidence c^v of each voxel can be aggregated into a global quality measure q^v, representing to which extent the location fulfils the suitability/confidence requirements of the decision-maker. For example, if according to the decision-maker the best locations are those with a high suitability and high confidence, q^v can be defined by $q^v = s^v c^v$. An alternative definition for $q^v$is $q^v = p\frac{s^v}{s^{max}} + (1-p)\frac{c^v}{c^{max}}$, where p denotes the relative importance of suitability as compared to confidence, s^{max} is the suitability degree of the best-suited location v' and c^{max} is the confidence of the location v'' associated with the highest confidence. Other definitions for q^v are possible.
- **One map using different visualization techniques.** One might also opt to represent both suitability and confidence in the same map, but using a different visualization technique for each of them. An option is to represent the areas of different confidence with differently sized voxels and use a coloured voxel fill to denote suitability. In Fig. 10, a TILES suitability and confidence model, using this option, is given. A part of the Hinderbanks region is shown. The suitability criteria are those for locating medium sand ('*P*(*medium sand*)'). Green denotes a high probability, red a lower probability. The larger the voxels are, the lower the data quality will be.

6 Practical Experiences from the TILES Project

The TILES project is a typical example of a real-life case study with imperfect information. Most of this information relates to lithological classification based on subsurface-sample descriptions.

Gathering information from the subsurface of the sea is expensive. As a direct consequence, samples are most commonly collected by various (volunteering)

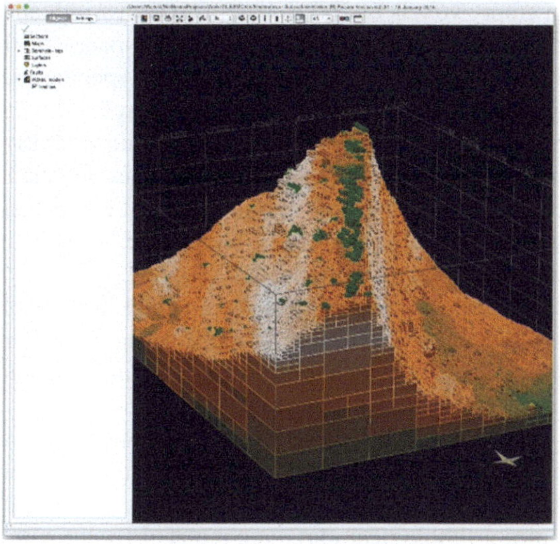

Fig. 10 Illustration of an extended suitability model for medium sand

parties in areas with planned infrastructure or other works (requiring detail at specific sites), for large-scale mapping (requiring coarse overall coverage), and for monitoring (requiring periodic sampling of a few points). For the entire region, this implies that some areas are better documented than others. A few areas are completely uncharted, which is the first complication when trying to construct a 3D geological subsurface model. As a secondary consequence of the acquisition cost, sites are not often revisited, meaning that new data and old data have to be used together. Due to constant water flow, however, the seabed is ever-changing. In many areas of the Belgian North Sea, the speed and impact of sand–wave migration, for example, is non-negligible. Removal of sand may expose older clay or gravel, and the addition of sand may cover clayey or gravelly areas. As a result, older information is inherently less trustworthy than newer observations.

Additionally, the method of sample collection and method of sample analysis or description has a large impact on the representativeness of the observation. Visual descriptions are less precise than laboratory techniques used to analyse sample characteristics, but commonly provide a more accurate measure of gravel content. Typically, newer samples are analysed with newer, more precise techniques.

The data quality aspects mentioned above are arguments for using an enriched attribute tree as depicted in Fig. 6. Indeed, this enriched attribute tree adequately reflects the semantics described above. However, practical constraints, inherent to the time-consuming, interactive construction process of a lithological model forced us to follow a much simpler, though less representative approach.

TILES use a fixed-resolution 3D voxel model reflecting all available information of the subsurface of the Belgian part of the North Sea. A predefined 3D space, encompassing the entire coastline, reaching some 20 km into the North Sea and ranging from 4 m above mean sea level to 75 m below it, is first meshed (resulting in empty voxels of 200 m length, 200 m width and 1 m height) and then filled with data. Voxels are in essence small subspaces of this region. Because the data are sparse and highly concentrated around specific locations, most voxels represent an area without samples. To fill these voxels with estimates of lithological class, the Sequential Indicator Simulation technique (SIS) is applied. SIS, based on the indicator kriging formalism, is a well-established method to simulate lithological class distributions which require modest computation time and is straightforward to implement [20, 21] and is, for example, used in the construction of 3D subsurface models of the Netherlands [22].

In SIS, data from samples contained in a voxel are considered to be 'accurate'. All remaining voxels are scanned using a random path. For each voxel, a neighbourhood is considered in which both hard data and previously interpolated data are used to estimate the target voxel's value.

The random nature of the path allows the construction of multiple, statistically equally probable 3D realizations of the model. From these realizations, probabilities of occurrence for each lithological class are calculated by dividing the number of times a lithological class is simulated by the number of simulations. From experiments, it was found that 50–100 simulations result in stable, representative results [22]. Finally, each voxel is assigned a "most likely" lithological class using the averaging method for indicator datasets described in [23].

Given the probabilities of occurrence, Shannon entropy can be used as an indicator of the variability throughout the different simulations [24]. If there is no value that clearly appears most, the entropy of the voxel will be high, reflecting a high degree of uncertainty that the predicted value is accurate. On the other hand, a voxel having the same value in almost every iteration will have low entropy. TILES use this entropy as an indicator of the interpolation-related part of (lithological) model uncertainty.

The generated geological voxel model is then further processed by geologists, who use their domain knowledge and expertise to correct detected anomalies. This is a labour-intensive and time-consuming process in which geologists can also adapt the precomputed model uncertainty of a voxel. A direct consequence is that, for the time being, preselection of sample data based on data quality aspects like sampling time, sampling method or analytical technique and reconstruction of the geological model based on this preselection only, is not possible without discarding the revisions made by experts, hence the current decision for using a simplified data quality approach in TILES, which only copes with (lithological) data-related model uncertainty. Further research aimed at maintaining the expert input is essential.

7 Conclusions and Further Research

In this chapter, we studied the need for an explicit data quality assessment in (geographic) multi-criteria decision-making. Aspects of the project Transnational and Integrated Long-term marine Exploitation Strategies (TILES) illustrate the importance of such an assessment. Data quality is an important, hitherto ignored aspect in decision-making related to sand and gravel extraction, and inadequate data quality can lead to misleading data-processing results, especially in cases where a lot of data stemming from different data sources (e.g., provided by volunteers) are involved. To overcome this problem, users should be provided with extra information regarding the propagation of data quality as included in the quality of the suitability results produced using these data.

A scientific solution for data quality assessment using (geographic) multi-criteria decision-making based on LSP suitability models relies on the identification and specification of elementary data quality aspects for the elementary attributes that are involved in the decision-making process. For each elementary data quality aspect, a corresponding elementary quality criterion is specified. Evaluation of these elementary quality criteria for a given voxel in a geographical voxel space yields elementary confidence scores for that voxel. These elementary confidence scores are then aggregated to an overall confidence degree by applying a novel, extended version of the LSP aggregation structure. Every LSP aggregator is extended so that it takes both suitability degrees and confidence degrees as arguments and computes a couple that consists of an aggregated suitability degree and an aggregated confidence degree. The aggregated suitability degree is computed in exactly the same way as in regular suitability modelling. For the aggregated confidence degree, a weighted sum of the input confidence degrees is computed, with the weights reflecting the impact of the corresponding input suitability degrees on the computation of the aggregated suitability degree. In our future research, we aim to study alternatives for the weighted sum aggregation in order to further improve the semantic properties of the aggregation, so that this aggregation even better reflects and complements expert reasoning.

Application of the proposed technique in the TILES project reveals that complex data-processing techniques like the one used to produce 3D geological models can put hard constraints on the full exploitability of the proposed technique for data quality assessment. Further research on more advanced 3D geological modelling should be aimed at minimization of mutual constraints of expert knowledge versus data-related uncertainty.

References

1. Belton, V., & Stewart, T. J. (2002). *Multiple Criteria Decision Analysis*. Dordrecht, The Netherlands: Kluwer Academic Publ.
2. Zeleny, M. (1982). *Multiple Criteria Decision Making*. New York, USA: McGraw Hill.
3. Jankowski, P., & Richard, L. (1994). Integration of GIS-based suitability analysis and multicriteria evaluation in a spatial decision support system for route selection. *Environment and Planning B, 21*(3), 326–339.
4. Rinner, C., & Raubal, M. (2004). Personalized Multi-Criteria Decision Strategies in Location-Based Decision Support. *Journal of Geographic Information Sciences, 10*(2), 149–156.

5. Kangas, A., Kangas, J., & Pykäkäinen, J. (2001). Outranking Methods as Tools in Strategic Natural Resources Planning. *Silva Fennica, 35*(2), 215–227.
6. Keeney, R. L. (1996). Value-focused thinking: identifying decision opportunities and creating alternatives. *European Journal of Operational Research, 92,* 537–549.
7. Janssen, R., & Rietveld, P. (1990). Multicriteria analysis and geographical information systems: an application to agricultural land use in the Netherlands. In H. J. Scholten & J. C. H. Stillwell (Eds.), *Geographical information systems for urban and regional planning* (pp. 129–139). Dordrecht, The Netherlands: Kluwer Academic Publ.
8. Banai, R. (1993). Fuzziness in geographic information systems: contributions from the analytic hierarchy process. *International Journal of Geographical Information Systems, 7*(4), 315–329.
9. Dujmović, J. J., De Tré, G., & Van de Weghe, N. (2010). LSP suitability maps. *Soft Computing, 14,* 421–434.
10. Triantaphyllou, E. (2000). *Multi-Criteria Decision Making Methods: A Comparative Study*. Dordrecht, The Netherlands: Kluwer Academic Publ.
11. Dujmović, J. J., & De Tré, G. (2011). Multicriteria Methods and Logic Aggregation in Suitability Maps. *International Journal of Intelligent Systems, 26*(10), 971–1001.
12. Van Lancker, V., Francken, F., Kint, L., Terseleer, N., Van den Eynde, D., De Mol, L., et al. (2017). Building a 4D Voxel-Based Decision Support System for a Sustainable Management of Marine Geological Resources. In P. Diviacco, A. Leadbetter, & H. Glaves (Eds.), *Oceanographic and Marine Cross-Domain Data Management for Sustainable Development* (pp. 224–252). Hershey, USA: IGI Global.
13. Kint, L., & Van Lancker, V. (2016). SediLITHO@SEA v2 (06/10/2016). Database lithological descriptions, with relevance to Belgian part of the North Sea. Brussels: Royal Belgian Institute of Natural Sciences (internal report).
14. Van Lancker, V. (2009) SediCURVE@SEA: a multiparameter sediment database, in support of environmental assessments at sea. In: Van Lancker V. et al. (eds.) Quantification of Erosion/Sedimentation patterns to Trace the natural versus anthropogenic sediment dynamics (QUEST4D). Final Report Phase 1. Science for Sustainable Development. Brussels: Belgian Science Policy 2009, 63p + Annexes.
15. van Heteren, S., & Van Lancker, V. (2015). Collaborative seabed-habitat mapping: uncertainty in sediment data as an obstacle in harmonization. In P. Diviacco, P. Fox, A. Leadbetter, & C. Pshenichny (Eds.), *Collaborative Knowledge in Scientific Research Networks* (pp. 154–176). Hershey, USA: IGI Global.
16. Dujmović, J. J. (2007). Preference Logic for System Evaluation. *IEEE Transactions on Fuzzy Systems, 15*(6), 1082–1099.
17. Zadeh, L. A. (1965). Fuzzy Sets. *Information and Control, 8,* 338–353.
18. Dujmović, J. J., & Larsen, H. L. (2007). Generalized Conjunction/Disjunction. *International Journal of Approximate Reasoning, 46*(3), 423–446.
19. De Mol, R., Tapia-Rosero, A., & De Tré, G. (2015) An Approach for Uncertainty Aggregation using Generalised Conjunction/Disjunction Aggregators. In: Proc. of the IFSA/EUSFLAT 2015 conference, pp. 1499-1506, Gijón, Spain.
20. Chilès, J.-P., & Delfiner, P. (2012). *Geostatistics – Modeling Spatial Uncertainty*. New Jersey, USA: Wiley & Sons.
21. Goovaerts, P. (1997). *Geostatistics for Natural Resources Evaluation*. New York, USA: Oxford University Press.
22. Stafleu, J., Maljers, D., Gunnink, J. L., Menkovic, A., & Busschers, F. S. (2011). 3D modelling of the shallow subsurface of Zeeland, the Netherlands. *Netherlands Journal of Geosciences, 90*(4), 293–310.
23. Soares, A. (1992). Geostatistical estimation of multi-phase structure. *Mathematical Geology, 24,* 149–160.
24. Wellmann, J. F., & Regenauer-Lieb, K. (2012). Uncertainties have a meaning: Information entropy as a quality measure for 3-D geological models. *Tectonophysics, 526–529,* 207–216.

VGI Imperfection in Citizen Science Projects and Its Representation and Retrieval Based on Fuzzy Ontologies and Level-Based Approximate Reasoning

Gloria Bordogna, Cristiano Fugazza and Alessandro Oggioni

Abstract The chapter investigates the kinds of imperfection affecting Volunteer Geographic Information (VGI) created by users eager to participate in some citizen science project. An approach based on the use of fuzzy domain ontologies and level-based approximate reasoning is suggested to represent and manage both the uncertainty of volunteers when describing their observations and the vagueness of ill-defined domain knowledge. This way one can model more reliable smart applications for creating VGI as well as can design less ambiguous spatial data infrastructures (SDIs) for sharing VGI with final stakeholders.

1 Introduction

"Citizen Science" nowadays indicates a collaborative practice to carry out scientific projects by the involvement of a large number of volunteer citizens who are called to perform some of the tasks [14]. This is an old practice; just think of the hundreds of millions of plants and animals which have been collected by volunteers over the years and which enrich the museums' collections. Nevertheless, thanks to the Internet and smart applications' diffusion on mobile devices, what changes with respect to the past is the sharp increase in the quantity, the timeliness, and the worldwide provenance of volunteers' contributions which make it possible to adopt this practice also for monitoring processes, thus constituting a new challenge for science. Most citizen science projects exploit Volunteer Geographic Information (VGI) [11], that is, volunteers are asked to provide information on various forms and nature, such as textual notes, pictures, measurements of properties relative to target objects, by associating a geographic reference with their observations [13]. Nevertheless, many researchers express criticism toward usability of VGI due to its unavoidable imperfection. For this reason, in [7], we analyzed the possible causes of VGI quality degradation by identifying some relevant dimensions of VGI that

G. Bordogna (✉) · C. Fugazza · A. Oggioni
IREA CNR, Via Bassini 15, Milan, Italy
e-mail: bordogna.g@irea.cnr.it

G. Bordogna and P. Carrara (eds.), *Mobile Information Systems Leveraging Volunteered Geographic Information for Earth Observation*, Earth Systems Data and Models 4, https://doi.org/10.1007/978-3-319-70878-2_10

may impact on quality assessment; mainly, critical points are the heterogeneity of expertise and commitment of the volunteers, the media formats of the various social media platforms that lead to a variety of data structures and heterogeneous semantics of data, the redundancy and sparseness of the generated content, and finally, the different intended purposes for which VGI was created and for which it will be used. A review of the policies adopted by citizen science projects for regulating, constraining, filtering, and correcting the data entered by volunteers so as to reduce as much as possible the uncertainty of their interpretation due to errors, ambiguities, incompleteness, and inaccuracy is reported in [7, 21]. We recall the main categories of approaches:

- ex-ante approaches have the objective of providing volunteers with easy-to-access resources for a correct creation of information and automatic mechanism for controlling data entry. These resources include templates with automatic error-checking capabilities to ease adoption of better data creation practices, controlled vocabularies, geographic gazetteers, and ontologies in the specific scientific domain of the project;
- ex-post approaches have the objective of cleansing and improving VGI once it is already created. A common approach is based on cross-referencing VGI with either other authoritative information from administrative and commercial data sets, such as land cover, land use, DEM, or gazetteers and ontologies, in the specific scientific domain [1];
- hybrid approaches which apply both ex-ante and ex-post methods.

In all the approaches, the use of gazetteers and ontologies is particularly suitable and diffused in citizen science projects that ask volunteers to identify objects of interest and to categorize them on the basis of the provided domain ontology. These projects require volunteers to contribute with information on the observation of a huge number of objects of interest that are geographically distributed (such as birds, plants) and their classification according to a related semantics (such as the classification into species). Observations come together with ancillary information, such as the area where the object is observed (geographic footprint), a description and photographs of the object, the time of the observation. These are the main characteristics of most citizen science projects in natural sciences, such as the many projects on bird and insect watching and plants observation. Many of these projects primarily require qualitative information, or ask for interpreting the meaning of each observed entity; for example, to understand the differences between subtle shades of feather colors or bird calls, or the ground pattern indicating potential archeological sites. Among the many bird-watching projects, we recall the *eBird*[1] project; *eBird* employs volunteers who contribute their observations to a worldwide database that helps scientists to learn more about the many species of birds, and the *House Finch Disease Survey*[2] project that assists scientists in tracking the spread of

[1]http://ebird.org/.

[2]http://birds.cornell.edu/hofi/.

bird diseases. Other interesting projects are the *Great Sun Flower* project[3] that gathers information about urban, suburban, and rural bee populations to learn about what is affecting pollination of gardens, crops, and wildlands. The *Wildlife Health Event Reporter* project[4] that is an experimental tool that records wildlife observations by citizens concerned about dead or sick wildlife. It is a part of the Wildlife Health Monitoring Network; a Web-based open-source system that supports data entry, storage, reporting, analysis, and exchange. These projects demand high precision when classifying the observed objects, while they may tolerate inaccuracies about the objects geolocation. For example, in the *inaturalist*[5] project, the tolerance of the geolocalization of the observed object can be defined by specifying an uncertainty radius around a point identified by geographic coordinates; in fact, it is not so relevant to have the exact position of a flying bird, or of a plant. On the other side, in order to achieve a high quality in the classification of the observed objects, a hybrid approach for quality improvement is adopted. Ontologies enriched with visual information in the form of pictures and drawings of the species are provided to help volunteers create correct classifications of observations, for which they have to upload a photo that is then submitted to the manual collaborative check and possible revision by the community.

Nevertheless, even with the support of ontologies, many volunteers still find it difficult to select the correct classification of the observations and many of them explicitly ask for help by the community when they feel unsure of their classification. This uncertainty can depend on the "vagueness" of the classification criteria, the low expertise of some volunteers, and the inadequate context of the observation. The direct experience we had in participating in the *Space4Agri*[6] project for the improvement of the agronomic sector in northern Italy motivated our proposal of defining a fuzzy ontology approach to support the ex-ante creation of in situ VGI about crop observations by means of a smart app. Within this project, we experienced both the uncertainty of volunteers who were asked to identify the crops and their actual crop growth stage when observing agronomic parcels in which crops may have grown with different speed from place to place within the same field, and the vagueness of the taxonomic classification of crops growth stages [5, 4].

In the chapter, we first introduce the problem of semantic interoperability and the causes of VGI imperfection. Then we introduce the ontologies and discuss their inadequacy for dealing with ill-defined knowledge. Thus, we introduce the notion of fuzzy ontology to represent ill-defined knowledge in a domain and level-based approximate reasoning to deal with the uncertainty of volunteers when creating VGI, by discussing a case study example.

---

[3]http://www.greatsunflower.org/.

[4]http://www.whmn.org/where/.

[5]http://www.inaturalist.org/.

[6]http://space4agri.irea.cnr.it/.

2 Imperfection in Semantic Interoperability of VGI

The management of VGI is characterized by the absence of standards for data deployment and retrieval on the Web. This deficiency has caused the creation of ad hoc approaches and, consequently, of great heterogeneity of VGI—both in contents, formats, and semantics. Nevertheless, this heterogeneity represents a collaborative view of reality complementary to that provided by scientific and authoritative geospatial data. Thus, VGI integration in SDIs would be of great usefulness [5, 18]. This issue has been addressed in the literature primarily by investigating syntactic interoperability in VGI sharing, by integrating existing heterogeneous VGI with respect to formats and services [21]. Semantic interoperability of VGI from different sources [6, 19] is generally disregarded in the literature. On the contrary, semantics is fundamental to understand the meaning of data since it defines the meaning of symbols in a language. In the last decades, the Semantic Web has developed methods and techniques to support semantic interoperability among applications. The cross-community interoperability initiative by OGC [8] provides recommendations to create, assemble, and disseminate VGI, specifically:

- the creation of ontologies that define the shared vocabulary used to model a domain with the definition of concepts, as well as their properties and relations;
- the extension of OGC services and standards to VGI deployment and access;
- the expansion of gazetteer functionalities so as to include conflation and semantic linking;
- the management of uncertainty and imprecision in VGI.

In order to enable semantic interoperability, [2] proposes a VGI semantic data model that defines the characteristics of VGI from distinct projects, i.e., applications, in terms of the possible categories of data that might be requested to volunteers. This VGI data model is intended to support the ex-post semantic integration of VGI when an application already exists. It provides the conceptual framework for the generation of shared descriptions of heterogeneous VGI. These descriptions should act as common interfaces to enable the querying and correct interpretation of the VGI provided by distinct projects through a single platform like in [7]. In order to define the VGI semantic data model, the authors in [2] review some VGI citizen science initiatives and classify them according to the type of contributions, namely VGI provided by sensor devices, georeferenced text, georeferenced features, i.e., abstraction of real-world phenomena such as objects (ex. buildings, POIs) and events. In their approach, to enable the accessibility of VGI, each VGI application must register in a service by defining an instance of the VGI data model through semantic annotation [17]. Semantic annotations must be performed manually since automatic processes are still missing. Thus, the subjectivity of the single annotator may force the interpretation of the original VGI, possibly created by multiple volunteers with heterogeneous characteristic.

Complementary approaches have been proposed that suggest applying ex-ante semantic mechanisms to support volunteers in creating VGI meaningful to the context of an application. Ex-ante semantic support for creating VGI has been recently addressed in two distinct ways: either by adopting the Semantic Web and linked data framework [13, 20, 22] or by relying on domain ontologies [4, 17]. To this end, they provide a contextual description of the meaningful concepts and relationship used in the application to support not only VGI authors but also VGI consumers. Forcing volunteers to select tags from a vocabulary of meaningful terms and relationships, and possibly complementing the tags with free text and pictures, can be considered as a kind of ex-ante semantic annotation directly performed by VGI authors. We believe that this approach is much more respectful of the original intended meaning of VGI than ex-post semantic annotation. Nevertheless, if the linked data approach can be suitable for analyzing VGI from multiple sources, linked data cannot be accessed and used when creating VGI from a mobile device, not connected to the Internet. This is the main rationale for adopting ontologies to support ex-ante semantic interoperability of VGI [4].

The problem of semantic interoperability of VGI is strictly related to the management of the imperfection that characterizes VGI. Four dimensions of VGI imperfection can be identified depending on the origin or cause of the imperfection itself:

- incomplete or inadequate knowledge of the volunteer participating in the project. This may be particularly impacting in citizen science projects involving as volunteers the general public, without providing sufficient training facilities and information on the scopes of the project;
- vagueness of the criteria provided to support volunteers in correctly describing or tagging their observations. This may happen in domains where volunteers are asked to tag the observed objects based on vague descriptions of category prototypes;
- limitations in the means of observation of volunteers. This may depend on the context of the observation, such as weather conditions, distance from the object, or means of observation and measurement, like poor human sight;
- uncertainty arising when analyzing VGI created by a consumer outside the project. This may be due to missing contextual information on the created VGI, such as the lack of appropriate metadata describing the VGI items.

Table 1 summarizes the results of an analysis relative to some citizen science projects based on the four dimensions defined above. It can be observed that most of the projects have an incomplete knowledge of the potential contributors, who can be citizens not necessarily experts in the field of the project. Most projects supply vague criteria to classify the objects of interest, mainly expressed linguistically, thus subject to the user's interpretation. About half of the projects require accurate observations, which can be hard to perform given that the major observation means used in these projects is the human sight. Finally, some projects release the collected VGI as open data, and thus, the potential users of VGI are unknown, and thus, some means to communicate and disambiguate the semantics of observations are needed.

Table 1 Imperfection of VGI within some citizen science projects

Projects' acronym	Incomplete knowledge of the VGI author	Vague classification criteria (expressed in the ontology)	Deficiency of observations' means	Missing knowledge of the VGI consumer
eBird (see footnote 1)	x	x	x	
House Finch Disease Survey (See footnote 2)		x	x	x
Great SunFlower (See footnote 3)	x			
Wildlife Health Event Reporter (See footnote 4)	x			
iNaturalist (See footnote 5)		x	x	
Space4Agri (See footnote 6)			x	x
Hai sentito il terremoto	x			
Atrapa el Tigre	x	x	x	
MIPP	x	x		
CSMON	x	x	x	x
Galaxy Zoo	x	x	x	x
Whale Song	x	x		
Secchi depth	x	x		
mPing	x	x		x
Loss of the Night	x	x		

http://www.haisentitoilterremoto.it/compile.html
http://www.atrapaeltigre.com
http://lifemipp.eu
https://www.csmon-life.eu
https://www.galaxyzoo.org
http://whalesong.net
http://www.secchidisk.org
http://www.nssl.noaa.gov/projects/ping
http://lossofthenight.blogspot.it

3 VGI Created by Means of Ontologies and the Problem of Representing Imperfection

In order for data to become unambiguous, meaningful, and useful, their semantic must be defined. In fact, data are raw, unorganized facts, or figures. When data are processed, organized, structured, and interpreted in a context, they become meaningful information.

When data are distributed or shared among various data producers and consumers, as it happens for georeferenced data created by the crowd and shared on the Web, it is necessary to have either a shared semantics through a common vocabulary to specify entities, concepts, and relationships (a domain ontology), or mapping rules between different local vocabularies (ontology alignment). To this aim, the role of ontologies is growing in significance with the growth of the social Web and the diversity of the user-generated contents. Ontology is a *formal* and *explicit* specification of a *shared conceptualization* [12]. It constitutes a common context model and plays a key role in inferring semantics behind user-generated contents:

- *conceptualization* refers to an abstract model of some part of the world which identifies the relevant concepts and relations;
- *explicit* means that the type of concepts, the relations between concepts, and constraints on their usage are explicitly defined;
- *formal* refers to the fact that ontology should be defined in a machine-readable form;
- *shared* means that the ontology should reflect the understanding of a community.

Specifically, in the Semantic Web context, ontologies provide a knowledge representation shared between agents, in order to understand each other in communication. Ontologies subsume taxonomies since besides the hierarchical organization of concepts through inheritance ("IsA") relationship, they represent other kinds of relationships, such as "part of."

There are several formal languages that can be used for expressing ontologies. Among these, in the following, we will introduce the Web Ontology Language (OWL) which is the W3C recommendation for coding ontologies for the Web.

3.1 OWL Ontologies and Reasoning

The Web Ontology Language (OWL) defines a family of formalisms for defining ontologies. Syntactically, OWL extends RDF with the primary aim of supporting the expressiveness and reasoning power of Description Logics (DLs) [16].

OWL allows for intensional definition of *classes* (*concepts* in the formalism of DL) and *properties*, that is, binary relationships between classes (named *roles* in DL). Classes and properties constitute the *Terminological Box* (TBox) of a knowledge base. OWL also allows for the definition of *instances* (*individuals*), the extensional definition of classes and properties, that constitute the *Assertional Box* (ABox). Finally, OWL allows for role-centric definition of modeling features (the RBox). In the following, we are going to refer to the original organization of the different "flavors" of OWL set by the W3C recommendation of February 2004 and not that defined in the OWL 2 recommendation [15]. Specifically, the recommendation defines three levels of extensions of OWL characterized by different trade-off of expressiveness and computational efficiency: OWL-Full, OWL-DLs, and OWL-Lite.

OWL-DLs can be viewed as machine-processable representation of Description Logics, suitable for interoperability and scalability of systems that promote reuse of data and reasoning infrastructure over the Web. OWL-DL is defined on the basis of the following symbols:

- N_c: Class names, equivalent to unary predicates like `Flower;`
- N_p: Property names, equivalent to binary predicates like `hasColour.Red`, `hasPetal_Number(> 20)`, etc.;
- N_i: Individual names, i.e., instances like `e1`, equivalent to constants;
- operators akin to Boolean ones, ($\sqcup$, $\sqcap$, $\neg$,), quantifiers ($\exists$, $\forall$), relational operators ($\geq$, $\leq$, =), the subsumption operator $\sqsubseteq$, etc.

Formally, an OWL ontology O consists of:

$$O = <\text{TBox, RBox, ABox}> \;=\; <N_c,\ N_p,\ N_i> \tag{1}$$

where the TBox contains axioms about classes, domain knowledge, describing relationships between concepts. For example: `Flower` $\sqsubseteq$ `Plant.`

The RBox contains axioms describing relations between role names, i.e., properties. For example: `hasPetal_Number(> 20)` $\sqsubseteq$ `hasPetal.`

The ABox contains axioms about instances (*facts*) describing relations between named individuals. For example: `sample1` $\neq$ `sample2.`

For example, to represent roses as flowers having more than 20 petals whose dimension is within 20 mm and 40 mm, we can write an axiom in TBox such as the following:

$$\begin{aligned}&\texttt{rose} \sqsubseteq \texttt{Flower} \sqcap (\exists\, \texttt{hasPetal\_Number}(>20)) \sqcap \\ &(\exists\, \texttt{hasPetal} - \texttt{Width}(\geq 20\text{mm} \sqcap \leq 40\text{mm}))\end{aligned} \tag{2}$$

A flower can either satisfy a predicate (i.e., it is true) or not (i.e., it is false). It cannot be simultaneously true and false since this would contradict the axiom of excluded middle of classical logic.

Finally, facts can be expressed by data axioms representing assertions, such as:

$$\texttt{Bird(Tweety)} \tag{3}$$

That states that *Tweety* is a bird.

The power of OWL-DL ontologies resides in the sound *inference* (a.k.a. *reasoning*) that can be computed on the basis of axioms in the knowledge base. Most importantly, DL inference is aimed at computing implicit information in the *open world* of the Semantic Web. This means that, when new assertions are added to the knowledge base, these are not allowed to disprove any previously inferred implications. Thus, a DL reasoner allows making inference for specific reasoning tasks that make OWL powerful for both knowledge modeling and processing. For example, a deductive reasoning task is aimed at exploiting the knowledge about a domain represented in the TBox in order to produce a classification of instances described in terms of their properties.

Let us assume that we have a TBox defined by axiom (2) and an ABox with the following facts about an instance x:

$$\texttt{Flower}(x),\ \texttt{hasPetalWidth} = 30\text{mm}(x), \texttt{hasPetal\_Number} = 40(x) \quad (4)$$

stating that x is a flower with petals width equal to 30 mm and with a number of petals equal to 40. By applying the axioms in the TBox (2), a DL reasoner can conclude that x is a rose:

$$\text{rose}(x) \quad (5)$$

3.2 *Inadequacy of OWL Ontologies to Represent Imperfect VGI*

OWL-DL allows for representing the world in terms of membership of instances to crisp concepts (sets) and binary relationships among instances. In particular, a given statement can be either true or false. As a consequence, OWL is suitable to modeling domains in which concepts or relationships have a precise definition and thus are not ill-defined, or vague/fuzzy by their very nature. Nevertheless, there are concepts of the real world and relationships that are intrinsically vague and fuzzy, due to their gradual nature. For instance, we may find difficulties in encoding as OWL-DL the ill-defined knowledge about Calla expressed as follows:

Calla is a *very large*, *long white* flower on *thick* stalks. (6)

Vague concepts, like “*very large,*” “*long,*” and “*thick,*” involve some fuzziness and vagueness for which a crisp and precise definition is impossible or does not make sense. What is the wideness of a flower that makes it “*very large*”? This is a matter of degrees depending on a subjective interpretation and, certainly, there is not a crisp transition between a Calla being “*very large*” and “*not very large*” that may be agreed upon by all possible volunteers.

Another possible source of imperfection in observations is when the volunteer is not completely sure about his/her observation. This may happen because he/she does not have adequate knowledge of the problem or because of deficiencies in the means of observation. This may also happen when the domain knowledge is encoded into a precise ontology. For example, consider the description of Rose provided by the axiom in definition (2). For an observer, it might be impossible to measure the width of petals precisely because he/she is far from the flower or does not have a ruler. In such situations, it might be questionable to state if the predicate `hasPetalWidth. (≥ 20 mm⊓ ≤ 40 mm)` is true or false.

Finally, there are situations that may involve both ill-defined knowledge and uncertainty of VGI authors. Let us consider this description of sparrows provided by Wikipedia:

> sparrows are plump *little* brown or *greyish* birds, *often* with black, yellow or white markings. Typically 10–20 centimetres long, they range in size from the chestnut sparrow (Passer eminibey), at 11.4 centimetres and 13.4 grams, to the parrot-billed sparrow (Passer gongonensis), at 18 centimetres and 42 grams. They have *strong, stubby* conical beaks with *decurved* culmens and *blunter* tips.

This description of sparrow types contains both fuzzy (identified by italic font) and precise predicates. It may be difficult for a bird-watcher to state if a bird is *little* and *greyish*. Besides, it might be even impossible to measure the actual length and weight of an observed bird or to estimate the measurements from a far observation point. Thus, it may be questionable whether to tag the observation as that of either a chestnut sparrow or a parrot-billed sparrow.

Fuzzy ontologies have been defined in order to encode the fuzziness of concepts when the domain knowledge is ill-defined [23].

3.3 Fuzzy Ontologies to Deal with Ill-Defined Domain Knowledge

In this paragraph, we introduce the fuzzy ontology to deal with ill-defined domain knowledge and precise observations. According to [23], a fuzzy ontology can be specified by using an extension of OWL-DL, specifically Fuzzy OWL-DL.

In order to introduce the basic concepts, we first need to recall the definition of a fuzzy set. Fuzzy sets have been introduced by Zadeh in 1965 [24] to deal with fuzzy concepts such as *low* temperature, *high* weight. Formally, a fuzzy set A with respect to a universe X is characterized by a membership function μ_A: X → [0, 1], assigning a membership degree, $\mu_A(x) \in [0,1]$, to each element x of the domain X.

$\mu_A(x)$ provides an estimation of the belonging of x to A. Typically, if $\mu_A(x) = 1$, it means that x definitely belongs to A, while if $\mu_A(x) = 0$, it means that x does not belong to A at all. $\mu_A(x) = 0.8$ means that x is partially an element of A, which could mean either that x does not satisfy all properties that characterize an element of A or that one lacks complete knowledge on x and cannot state precisely if it is an element of A.

Accordingly, in fuzzy logics, the notion of degree of membership, $\mu_A(x)$, of an element $x \in X$ to the fuzzy set A over X is regarded as the degree of truth in [0,1] of the statement "x is A." This interpretation is applied in fuzzy-DL, where a concept A, rather than being interpreted as a classical set, is interpreted as a fuzzy set, and thus, concepts can be fuzzy. As a consequence, the statement "a is A," i.e., a: A, has a degree of truth in [0,1] given by $\mu_A(a)$, the degree of membership of the individual element "a" to the fuzzy set A.

Boolean operators defined to combine classic sets have been generalized to combine fuzzy sets so that the logic *intersection, union,* and *complement* are, respectively, defined by a *tnorm* (min), a *tconorm* (max), and *not* (1-) operators.

$$(\mathrm{A} \wedge \mathrm{B})(\mathrm{x}) = \mathrm{tnorm}(\mathrm{A}(\mathrm{x}),\ \mathrm{B}(\mathrm{x})) = \min(\mu_{\mathrm{A}}(\mathrm{x}),\ \mu_{\mathrm{B}}(\mathrm{x})) \quad (7)$$

$$(\mathrm{A} \vee \mathrm{B})(\mathrm{x}) = \mathrm{tconorm}(\mathrm{A}(\mathrm{x}),\ \mathrm{B}(\mathrm{x})) = \max(\mu_{\mathrm{A}}(\mathrm{x}),\ \mu_{\mathrm{B}}(\mathrm{x})) \quad (8)$$

$$\neg \mathrm{A} = \mathrm{X} - \mathrm{A}(\mathrm{x}) = 1 - \mu_{\mathrm{A}}(\mathrm{x}) \quad (9)$$

Another important operator used to model the `IsA` relationship of specialization/generalization between concepts in a hierarchy is the *inclusion between fuzzy sets*, denoted by $\rightarrow$, defined as follows:

$$\mathrm{A} \rightarrow \mathrm{B} = \inf_{\mathrm{x} \in \mathrm{X}} i(\mathrm{A}(\mathrm{x}), \mathrm{B}(\mathrm{x})) \quad (10)$$

in which *i* denotes a fuzzy implication. Several definitions of the fuzzy implication *i* have been provided with distinct interpretations [6]; the most used one is the *Mamdani* fuzzy implication used in control systems and defined as follows:

$$i(\mathrm{A}(\mathrm{x}), \mathrm{B}(\mathrm{x})) = \min(\mu_{\mathrm{A}}(\mathrm{x}),\ \mu_{\mathrm{B}}(\mathrm{x})) \quad (11)$$

Another definition is the *Dienes* implication generalizing the classic implication definition:

$$i(\mathrm{A}(\mathrm{x}),\ \mathrm{B}(\mathrm{x})) = \neg\, \mathrm{A}(\mathrm{x}) \vee \mathrm{B}(\mathrm{x}) = \max(1 - \mu_{\mathrm{A}}(\mathrm{x}),\ \mu_{\mathrm{B}}(\mathrm{x})) \quad (12)$$

Fuzzy implications can be used to determine the degree of the subset relationship of a fuzzy subset A in the fuzzy set B, both defined over X: the degree of subsumption of A in B, i.e., A is a more specific concept than B, is denoted by $\mathrm{A} \subseteq \mathrm{B}$.

In fact, note that, if $\forall \mathrm{x} \in [0,\ 1]$ $i(\mathrm{A}(\mathrm{x}),\ \mathrm{B}(\mathrm{x})) = 1$ holds, then $\forall \mathrm{x} \in [0,1]$ $\mathrm{A}(\mathrm{x}) \le \mathrm{B}(\mathrm{x})$ holds too, i.e., A is perfectly included in B, and $\mathrm{A} \rightarrow \mathrm{B} = 1$. Of course, it may be that $\mathrm{A} \rightarrow \mathrm{B} = \mathrm{v} < 1$ when x is not a full member of both A and B.

At this point, we can represent an ill-defined statement like (6) describing a `calla` by the following fuzzy axiom:

$$\texttt{calla} \sqsubseteq \texttt{Flower} \sqcap (\exists\, \texttt{hasSize.very\_Large})) \sqcap (\exists\, \texttt{hasPetalWidth.Long}) \sqcap (\exists\, \texttt{hasColour.White}) \sqcap (\exists\, \texttt{hasStalks.Thick}) \quad (13)$$

where

- `Flower` is a crisp predicate so that $\mu_{\texttt{Flower}}(\mathrm{x}) = 1$ if x is a `flower`; otherwise, $\mu_{\texttt{Flower}}(\mathrm{x}) = 0$.
- `hasSize.very_Large`, `hasPetalWidth.Long`, `hasColour.White`, and `hasStalks.Thick` are fuzzy predicates represented by fuzzy

sets with membership functions $\mu_{\texttt{Size.very\_Large}}$, $\mu_{\texttt{PetalWidth.Long}}$, $\mu_{\texttt{Colour.White}}$, $\mu_{\texttt{Stalks.Thick}}$. For the sake of simplicity notice that, in practical contexts, we can use trapezoidal-shaped membership functions which can be simply defined by formula (14): trapezoidal functions can be uniquely specified by a tuple ($a \le b \le c \le d$) where *a,b,c,d* are values defined on the basic numeric domain of a property *P* like *Size*, *PetalWidth,* and *Stalk*, respectively.

$$\mu_P(x) = \begin{cases} 0 & x \le a \, or \, x \ge d \\ \frac{x-a}{b-a} & a < x < b \\ \frac{d-x}{d-c} & c < x < d \\ 1 & b \le x \le c \end{cases} \tag{14}$$

For short in formula (14), we replaced the instance property numeric value p_x with x.

- $\sqcap$ is the intersection operator between fuzzy sets, $\wedge$, defined by the minimum as in (7), and $\sqsubseteq$ is the subsumption operator between fuzzy sets and defined by the Mamdani fuzzy implication *i,* the minimum in (11).

Let us assume that the VGI author can provide a precise observation of a flower x by measuring the size, the petal width, and the type of stalk.

Given these precise measurements and the membership functions $\mu_{\texttt{Size.very\_Large}}$, $\mu_{\texttt{PetalWidth.Long}}$, $\mu_{\texttt{Colour.White}}$, $\mu_{\texttt{Stalks.Thick}}$ defining the meaning of the fuzzy predicates, we can compute their degrees of satisfaction, and finally, the truth of "x is a `calla`" by applying approximate reasoning based on fuzzy predicates [3]:

$$c\mu_{\text{calla}}(\text{x}) = \min(1, \mu_{\text{Size.very\_Large}}(\text{x}), \mu_{\text{PetalWidth.Long}}(\text{x}), \mu_{\text{Colour.White}}(\text{x}), \mu_{\text{Stalks.Thick}}(\text{x})) \tag{15}$$

Now, let us assume that we have another axiom in our fuzzy TBox defining a `magnolia` flower as follows:

$$\texttt{magnolia} \sqsubseteq \texttt{Flower} \sqcap (\exists\, \texttt{hasSize.very\_Large}) \sqcap (\exists\, \texttt{hasPetal} - \texttt{Width.Long}) \sqcap (\exists\, \texttt{hasColour.White}) \sqcap (\exists\, \texttt{hasStalks.Woody}) \tag{16}$$

The same instance x satisfying axiom (13) can satisfy the axiom (16) to a not-null degree as defined in the following (17) formula:

$$\mu_{\text{magnolia}}(\text{x}) = \min(1,\ \mu_{\text{Size.Large}}(\text{x}),\ \mu_{\text{PetalWidth.Long}}(\text{x}),\ \mu_{\text{Colour.White}}(\text{x}),\ \mu_{\text{Stalks.Woody}}(\text{x})) \tag{17}$$

This way, we have that an instance x is associated with two distinct kinds of flower to distinct degrees, thus representing the uncertainty of the classification.

Now, let us assume that we have five axioms in our fuzzy TBox, (16) and (17) defining calla and magnolia, and the further axioms in (18) defining the exclusive IsA relationship between angiospermae and calla, magnolia and rose:

$$\begin{gathered}\texttt{rose} \sqcap \texttt{magnolia} \sqcap \texttt{calla} \equiv \bot \; \texttt{calla} \sqsubseteq \texttt{angiospermae} \sqcap \\ \texttt{magnolia} \sqsubseteq \texttt{angiospermae} \sqcap \texttt{rose} \sqsubseteq \texttt{angiospermae}\end{gathered} \tag{18}$$

By knowing the degrees to which the flower is a magnolia and a calla, for example, 0.8 and 0.6, respectively, we can obtain the degree to which x is a rose by applying the following approximate reasoning deduction rule in formula (19) [3]:

$$\begin{aligned}&\texttt{rose} \sqsubseteq \texttt{angiosperme} \sqcap \neg\, \texttt{calla} \sqcap \neg\, \texttt{magnolia}\\ &\mu_{\texttt{rose}}(\mathrm{x}) = \min(\texttt{angiospermae}(\mathrm{x}), \mathrm{Not}\; \texttt{calla}(\mathrm{x}), \mathrm{Not}\; \texttt{magnolia}(\mathrm{x}))\\ &= \min(1,\ 1-0.8;\ 1-0.6) = 0.2\end{aligned} \tag{19}$$

If the degrees to which the observed flower is a calla or a magnolia are both *very low*, for example, 0.1 and 0.2, the possibility that the flower is a Rose is *high*, equal to 0.8.

If we add to the fuzzy TBox also axiom (2) so that a rose can be defined by either axiom (2) or axiom (19) and assume that x satisfies axiom (2) to the degree 0.4, we can conclude that x is a rose to the degree obtained by applying the deduction formula (20):

$$\texttt{rose} \sqsubseteq (2) \sqcup (19).$$

$$\mu_{\texttt{rose}}(\mathrm{x}) = \max(0.4,\ 0.2) = 0.4 \tag{20}$$

Finally, fuzzy quantifiers can be applied to inclusion axioms as well, allowing to express vague quantified expressions, for instance, that “most birds fly” as follows:

$$(\mathit{Most})\texttt{Bird} \sqcap \texttt{FlyingObject} \tag{21}$$

Here, the fuzzy quantifier *Most*, defined by a fuzzy set with a monotonic not decreasing membership function on the domain of natural numbers, replaces the classical universal quantifier ∀ assumed in the inclusion axioms so that the above expression can be satisfied to a degree in the case in which the set of FlyingObjects contains exceptions, i.e., birds that do not fly.

4 Modeling Uncertainty of VGI Authors Using a Fuzzy Ontology

Nevertheless, Fuzzy-DL reasoning illustrated so far is suitable to deal with precise observations and ill-defined knowledge. A dual situation that may happen in the real world of observations is when the volunteer is not completely sure about his/her observation, either because he/she does not have adequate knowledge of the terms or because of deficiencies of the means of observation. This may happen when the domain knowledge is encoded into both a precise ontology and a fuzzy ontology. In [4], a real case study in agriculture is described in which the need to represent and manage the uncertainty of the observation when creating VGI based on a classic ontology has emerged.

In this chapter, we will discuss a simple case study considering the following descriptions of the three main groups of roses provided by Wikipedia and represented by the fuzzy ontology depicted in Fig. 1:

- "Wild roses are mainly reddish, 4–6 cm diameter, generally with 4–5 petals."
- "Old Garden roses are notably fragrant, double-flowered blooms primarily in shades of white, pink and crimson-red."
- "Modern roses are well-formed with large, high-centred buds, and their colors range from shades of deep yellow, apricot, copper, orange, true scarlet, yellow bi-colors, lavender, gray, and even brown were now possible."

It can be noticed that some fuzzy characteristics may be difficult to observe or perceive: for example, the fragrance, besides being a subjective property, it is also difficult to appreciate from a far observation point, thus one can be uncertain in specifying *scented*, *strongly scented*, or *fragrant*.

There is often a misunderstanding on the modeling of uncertainty and imprecision or fuzziness [23]. Uncertainty is relative to the truth of a proposition, while imprecision/fuzziness is relative to the unspecificity of the proposition with respect to the possible worlds. Under uncertainty theory, statements are either completely

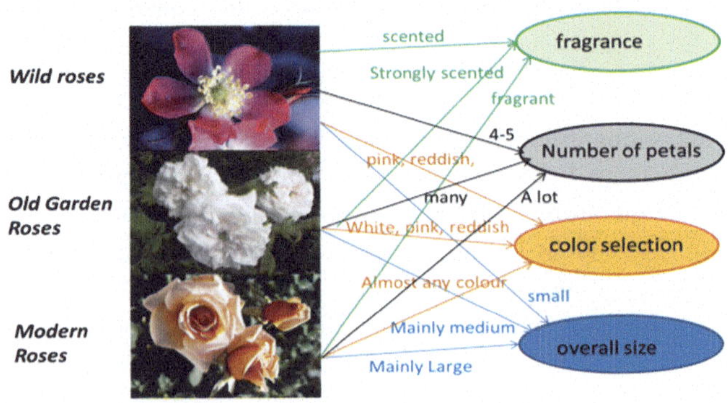

Fig. 1 Fuzzy ontology of the three main groups of roses

true or completely false but we do not know if they are true of false, so we define a probability or possibility distribution over the worlds. For example, the statement "x is a Flower" is a crisp one: x can be either a flower or not, it cannot be partially a flower, the degree that we can associate with this statement is relative to our knowledge on the truth about x being a flower, which may depend on some deficiency of the observation.

Imprecision/fuzziness of a proposition are modeled by fuzzy concepts represented by fuzzy sets defined on the domain of possible worlds so that the more the values of the domain that have a not-null membership degree to the fuzzy set, the more the concept is unspecific, i.e., imprecise/fuzzy.

We can model observations affected by some deficiency by alternative statements; for example, by observing a rose from a far point of view, one could either specify the uncertainty on the truth of a precise predicate such as:

(a) "*I am 0.4 uncertain that the Petals are 5*"

or one could express a certain fuzzy predicate such as:

(b) "*I am sure that x has a few Petals*"

or both an uncertain and fuzzy predicate such as:

(c) "*I am 0.2 uncertain that the Petals are 4 or 5*".

In all these statements, we can notice that the uncertainty degrees (0.4, 0.2, and 0 in statements (a), (c), and (b), respectively) are decreasing and are inversely related to the amount of imprecision/fuzziness, i.e., unspecificity, of the predicates in the same statements (*5 petals* is specific, while the unspecificity of *4–5 petals* is lower than that of *a few petals*). One can guess that the total amounts of uncertainty plus imprecision/fuzziness in all statements describing the same observation (the overall imperfection) are constant and depend on the degree of overall deficiency of the observation so that the greater the deficiency the greater is the total amount of uncertainty plus imprecision/fuzziness.

In fuzzy databases, the occurrence of both uncertainty and imprecise/fuzzy (unspecific) values FV have been modeled by combining the uncertainty u with the membership function μ_{FV} of the fuzzy value so as to derive a modified membership function $\mu'_{FV}(x) = i(u, \mu_{FV}(x))$, where i is an implication function like the ones defined in formulae (10) and (11) whose effect is either to reduce or to expand the imprecision/fuzziness (unspecificity) of the fuzzy value FV based on the uncertainty degree u. As pointed out in [10], although these proposals can be useful in many applications, unfortunately, they are inappropriate when reasoning in fuzzy databases. In fact, it is reasonable to think that a small uncertainty on a value means that one cannot exclude as possible some near values, but still one can exclude as possible the far values, i.e., very dissimilar values, from FV [6, 10]. For this reason, we defined a novel approximate reasoning model when retrieving VGI affected by both uncertainty and fuzziness. Our proposal is based on the following assumptions:

- volunteers can create VGI with the support of a fuzzy ontology, from where they can select simple concepts, which are discriminating characteristic (property values) of the instances they have to tag, such as *fragrant*, *a few petals, color White, size 30 cm*, etc., which can be fuzzy;
- each property of an instance can be tagged only by a single value, possibly fuzzy, selected from the fuzzy ontology;
- each selected (possibly fuzzy) value can be associated with a degree *d* in [0,1] representing the overall deficiency of the observation, i.e., equivalent to the overall imperfection of VGI, which may be due to both the limitations of the means of observation (far point of view, low resolution of the means of observation) and/or the subjective inexperience of the volunteer who may find it difficult to interpret the meaning of the fuzzy values.

All (imperfect) VGI items are stored in a geodatabase that is made available for querying to potential VGI consumers, users. Users' queries are expressed by selecting concepts from the fuzzy ontology and by specifying a maximum level of tolerable defect *d*. An example of query can be "retrieve VGI observations of Wild Roses with maximum defect $d < 0.3$." Thus, when answering user queries we must take into account the degree of defect *d* as described in the following.

Hereafter, we will make an example relative to the fuzzy ontology describing `Roses` and their following observable development stages [9]: `Visible Petals (VP)`, `Open Flower (OF)`, and `Senescing Flower (SF)` reported in Fig. 2a. The fuzzy development stages `F-VP`, `F-OF`, and `F-SF` were defined by exploiting the fact that there is a gradual transition of the characteristics of each stage to the next one. The triangular membership functions were chosen for their simplicity and point-like nucleus, nevertheless, we could use other definitions. A volunteer can tag his/her observation of a flower as a rose with one of the development stages above and by providing the uncertainty or defect of the observation:

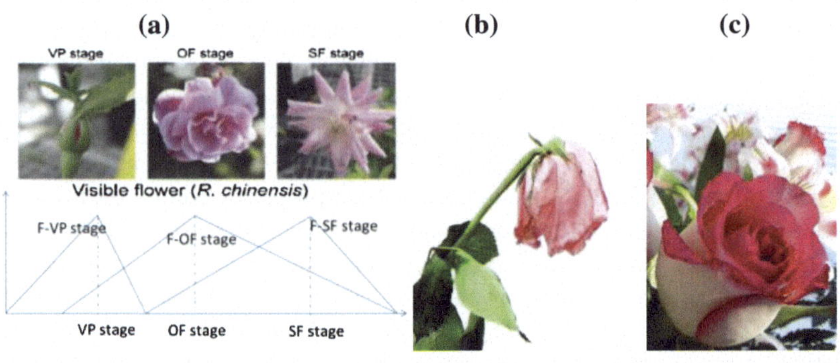

Fig. 2 **a** Fuzzy development stages of roses, and rose samples of questionable stage (**b**) and fuzzy color (**c**)

- $d = 1$ means maximally uncertain and imprecise description, then the specified property value, for example, `F-VP`, must be interpreted with maximum fuzziness, i.e., more stages close to the selected one are possible to a distinct degree. See the case depicted in Fig. 2b, where it is questionable to state if the rose in the middle is in the `VP` or `SF` stage. As far as categorical values, for example, colors, indeed, one can be inaccurate in selecting one value in a domain like {*white, pink, yellow, red, any color*} to tag the rose in Fig. 2c, so one can cope with the inaccuracy by associating a degree expressing the imprecision of the selection;
- $d = 0$ means no defect; the most certain and most precise description one can make as far as the possible choices in the fuzzy ontology. For example, the specified development stage `VP` is the only possible;
- by increasing *d* toward 1 it means that the description becomes more and more imperfect, and thus, the selected stage or the color is not exclusive.

To model this behavior, *d* defines a threshold on the membership function of the selected fuzzy value so that only the (fuzzy) values compatible with the selected one to a degree greater than $(1 - d)$, are considered as possible values of the observation. When the fuzzy concept has a point-like core, in the case of $d = 0$, only one precise value is possible. Nevertheless, when the fuzzy concept has a not punctual core (see the concepts *many, fragrant, scented* in Fig. 3), more values for the property are still possible, i.e., all the values in the core set.

Formally, let us consider an observation of an instance x defined by a pair $(\mathrm{FV}, d)_{\mathrm{x}}$ in which FV is a simple (fuzzy) concept of the fuzzy ontology *O*, with basic domain D, and *d* in [0,1] is the defect in stating FV as the observed value of a property of the instance *x*. $(\mathrm{FV}, d)_{\mathrm{x}}$ indicates that the volunteer is uncertain to the degree *d* on the truth of "x is FV." We translate it into the fuzzy axiom:

"*x* is FV with a certainty at most equal to $(1 - d)$": $Certain(x \text{ is FV}) \leq 1 - d$.

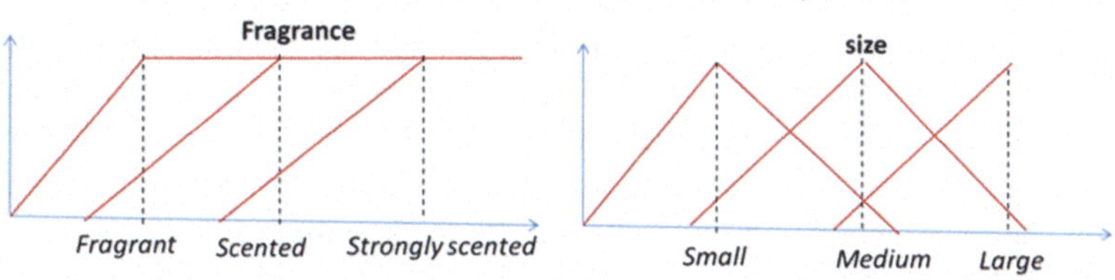

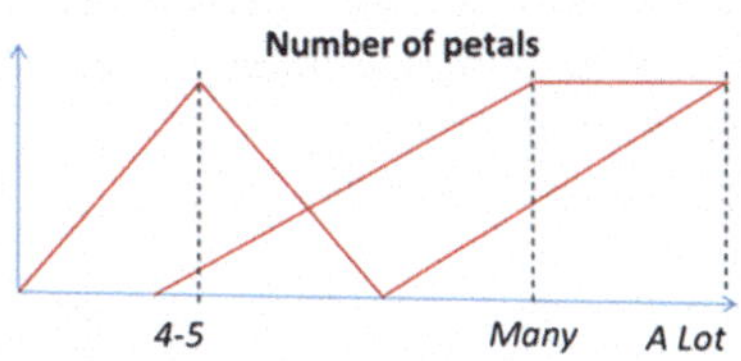

Color

	white	yellow	pink	reddish	Any color
white	1	0.2	0.6	0.4	1
yellow	0.2	1	0	0.3	1
pink	0.6	0	1	0.7	1
reddish	0.4	0.3	0.7	1	1
Any color	1	1	1	1	1

Fig. 3 Membership functions of the fuzzy values of fragrance, size, and number of petals, and possibility of co-occurrence of colors

Based on this axiom and on the membership function of FV, we can compute the degree of possibility π (x is *V*) that "*x* is *V*" is true, i.e., for each concept *V* defined on the basic domain D of FV, and belonging to the fuzzy ontology *O*:

Given *Certainty*(x is FV) $\leq 1-d$, then $\forall V \in$ D and $V \sqsubseteq O$ "x is *V*" with possibility degree:

$$\pi(\mathrm{x\,is}\,V) = \mu_V(\mathrm{x}) = \mu_{FV}(V) \quad \text{if} \quad \mu_{FV}(V) \geq 1 - d, \text{ else } \mu_V(\mathrm{x}) = 0 \tag{22}$$

When $d = 0$, precise and certain observation, $\mu_V(\mathrm{x}) = 1$ only if $\mu_{FV}(\mathrm{V}) = 1$ and *V* is defined in the fuzzy ontology *O*. When $d = 1$, maximally imprecise and uncertain description, $\mu_V(\mathrm{x}) > 0 \quad \forall\ V \sqsubseteq O$ and $V \in Support(FV) \subseteq$ D. By increasing *d*, more concepts of the fuzzy ontology become possible.

$$\begin{array}{c}\forall\, d_1 < d_2 \in [0,1],\ \{V_1, \ldots V_n \mid \mu_{FV}(V_i) \geq 1 - d_1\} \subseteq \\ \{V_1, \ldots V_m \mid \mu_{FV}(V_i) \geq 1 - d_2\}\end{array} \tag{23}$$

Notice that, when the membership function μ_{FV} defines a crisp set on the domain D, i.e., $\mu_{FV}(\mathrm{x}) \in \{0,1\}$, formula (22) reduces to the following:

$$\pi(\mathrm{x\,is}\,V) = \mu_V(\mathrm{x}) = 1 \quad \text{if} \quad \mu_{FV}(V) \geq 1 - d, \text{ else } \mu_V(\mathrm{x}) = 0 \tag{24}$$

In this case, when all concepts have crisp membership functions, the ontology is a classic one and the reasoning reduces to classic reasoning.

4.1 Example of Retrieval of Imperfect VGI Based on a Fuzzy Ontology and Level-Based Reasoning

Let us make a simple example. Assume the definition of the membership functions of the roses' fuzzy development stages in Fig. 2a, and the membership functions of the properties *fragrance, size, number of petals,* and *color* in Fig. 3. Specifically, an entry of the matrix of colors indicates the possibility of co-occurrence of the two colors in a single rose by considering all varieties. Thus, the possibility of finding roses with shades of white and yellow is 0.2, for pink–white roses is 0.6, and so on.

Assume the following three fuzzy axioms in TBox in Table 2 where each axiom admits a degree of possibility in [0,1] to be true, defining the three groups of roses, and let us define the intersection and inclusion by the min.

Observing the rose *x* in Fig. 2c, we can describe it as follows:

$$\begin{array}{l} x \sqsubseteq (\exists\, \mathtt{rose}, d_1 = 0) \sqcap (\exists\, \mathtt{hasFragrance.fragrant}, d_2 = 1) \sqcap (\exists\, \mathtt{hasPetals.many}, d_3 = 0.1) \sqcap \\ (\exists\, \mathtt{hasColor.pink}, d_4 = 0.5) \sqcap (\exists\, \mathtt{hasSize.large}\ , d_5 = 0.9)) \sqcap (\exists\ \mathtt{hasStage.F\_VP}, d_5 = 0.8) \end{array} \tag{25}$$

Table 2 TBox defining the three main types of Roses

1	Wild-rose ⊑ rose ⊓ (∃hasFragrance.scented) ⊓ (∃hasPetals.4-5) ⊓ (∃hasColor.pink-reddish) ⊓ (∃hasSize.small)
2	Old-Garden-rose ⊑ rose ⊓ (∃hasFragrance.strongly-scented) ⊓ (∃hasPetals.many) ⊓ (∃ hasColor.white-pink-reddish) ⊓ (∃Size.medium)
3	Modern-rose ⊑ rose ⊓ (∃hasFragrance.fragrant) ⊓ (∃hasPetals.a-lot) ⊓ (∃ hasColor.any) ⊓ (∃ hasSize.large)

In the geodatabase, we store ($Certainty$(rose)$_x$ = 1) ($Certainty$(hasFragrance.fragrant)$_x$ = 0), ($Certainty$(hasPetals.many)$_x$ $\leq$ 0.9), ($Certainty$(hasColor.pink)$_x$ $\leq$ 0.5), ($Certainty$(hasSize.large)$_x$ $\leq$ 0.1), ($Certainty$(hasStage.F_VP)$_x$ $\leq$ 0.2) together with the date and time of the observation (25 May 2016, 12:00) as well as its geographic coordinates.

Let us assume that a user wants to retrieve from the VGI collection all observations of Modern-roses that are blossomed, i.e., in Open Flower stage (OF), observed before 30 May 2016. The query is translated into the following:

$$x \sqsubseteq (\exists \text{ HasDate} < 30 \text{ May } 2016) \sqcap (\exists \text{ Modern-rose}) \sqcap (\exists \text{ hasStage.OF}) \tag{26}$$

We can compute the degrees of possibility that x satisfies the query as follows:

$$\min(\mu_{\text{date}<30\text{MAY}}(x.Date.05/25/2016/12:00), \mu_{\text{Modern-rose}}(x), \mu_{\text{OF}}(x.Stage.F\_VP))$$

where $\mu_{\text{Modern-rose}}$ (x) is computed by applying axiom 3 of the fuzzy TBox in Table 2 as follows:

$$\begin{aligned}\mu_{\text{Modern-rose}}(\text{x}) =& \min(\mu_{\text{rose}}(\text{x}), \mu_{\text{fragrant}}(\text{x.Fragrance.fragrant}), \mu_{\text{a-lot}}(\text{x.Petals.many}),\\ & \mu_{\text{any}}(\text{x.Color.pink}), \mu_{\text{large}}(\text{x.Size.large}))\\ =& \min(1,1,1,1,1) = 1.\end{aligned}$$

Furthermore
$\mu_{\text{OF}}(\text{x.Stage.F}_{\text{VP}}) = \mu_{\text{F-VP}}(\text{OF}) > 0$ since $\mu_{\text{F-VP}}(\text{OF}) = 0.5 \geq 1 - 0.8$. Thus, we can rank instance x based on its possibility to satisfy the query:

$$Rank(x) = \min(1, 1, \mu_{\text{F-VP}}(\text{OF})) = \min(1, 1, 0.5) = 0.5$$

A linguistic summary of the observation could be expressed as follows: "it is *fully possible* that x is a Modern rose but it is only half possible that x is in the Open Flower stage."

Assuming the following query asking for Old-Garden-roses :

$$x \sqsubseteq (\exists \text{ hasDate} < 30 \text{ May } 2016) \sqcap (\exists \text{Old-Garden-rose})\ ) \tag{27}$$

we should evaluate axiom 2 in Table 2 as follows:

$$\mu_{\text{Old Garden rose}}(x) = min(\mu_{rose}(x), \mu_{\text{strongly-scented}}(x.Fragrance.fragrant), \\ \mu_{\text{many}}(x.Petals.many), \mu_{\text{white-pink-reddish-color}} \\ (x.Color.pink), \mu_{\text{medium}}(x.Size.large)) = min(\ 1, 1, 1, 1, 0.2\ ) = 0.2$$

In this case, *x* would appear among the results of the query but with a lower rank with respect to rank computed for `Modern-roses`. Thus, we can conclude that "it is fully possible that *x* is a `Modern-rose` but there is a low possibility that it is an `Old-Garden-rose` too."

Let us assume that we observe a rose *x* without defects (for each fuzzy predicate *d* = 0).

$$x \sqsubseteq (\exists\ \texttt{rose}, d_1 = 0) \sqcap (\exists\ \texttt{hasFragrance.fragrant}, d_2 = 0) \sqcap \\ (\exists\ \texttt{hasPetals.many}, d_3 = 0) \sqcap (\exists\texttt{hasColor.pink}, \\ d_4 = 0) \sqcap (\exists\ \texttt{hasSize.large}, d_5 = 0)) \sqcap (\exists\texttt{F\_hasStage.F\_VP}, d_5 = 0) \tag{28}$$

In this case, in the geodatabase, we store (*Certainty*(`rose`)$_x$ = 1) (*Certainty* (`hasFragrance.fragrant`)$_x$ = 1), (*Certainty*(`hasPetals.many`)$_x$ = 1), (*Certainty*(`hasColor.pink`)$_x$ = 1), (*Certainty*(`hasSize.large`)$_x$ = 1), (*Certainty*(`hasStage.F_VP`)$_x$ = 1) together with the date and time of the observation (`25 May 2016, 12:00`) as well as its geographic coordinates.

By asking the query defined in formula (26), we would not retrieve *x*. In fact, $\mu_{OF}(\text{x.Stage.F}_{\text{VP}}) = \mu_{\text{F-VP}}(\text{OF}) = 0$ since $\mu_{\text{F-VP}}(\text{OF}) = 0.5 \leq 1 - 0$.

On the other side, when *d* = 1 (worst situation), we are in the framework of approximate reasoning with a fuzzy ontology, since the threshold in formula (22) is at level zero (1−*d*) = 0, and thus, it does not influence the results. Thus, in this framework, the fuzzy ontology represents the ill-defined knowledge assuming the worst observation conditions. The defect degree is used to limit the imprecision/fuzziness (unspecificity) by which we interpret the concepts in the fuzzy ontology.

5 Conclusions

In this chapter, we analyzed several aspects of uncertainty affecting VGI in citizen science projects. More closely, we introduced the notions of OWL ontology and Fuzzy OWL ontology; the latter employed to cope with ill-defined knowledge. Finally, we proposed an extension of the fuzzy ontology approach to model uncertainty of the volunteer when creating VGI due to deficiencies of the observation. The adoption of fuzzy ontology with level-based reasoning offers several advantages with respect to crisp definitions:

- we can model both precise and uncertain creation of VGI, so coping with the limitations of the observation means and context;
- we support inexperienced volunteers who are unable to interpret the meaning of some concepts in the ontology by allowing them to select "vague," i.e., fuzzy, concepts;
- when analyzing the created VGI users, we can filter VGI reports based on maximum levels of imperfection of VGI they can tolerate for their application needs.

Acknowledgements The present work was partially supported by the FHfFC project jointly funded by CNR and Regione Lombardia (Accordo Quadro di collaborazione tra Regione Lombardia e il Consiglio Nazionale delle Ricerche (CNR) D.G.R. n. 3866, 17/07/2015).

References

1. Al-Bakri, M., & Fairbairn, D. (2010). Assessing the accuracy of 'crowd sourced' data and its integration with official spatial data sets. In N.J. Tate, & P.F. Fisher (Eds.), Proceedings of the Ninth International Symposium on Spatial Accuracy Assessment in Natural Resources and Environmental Sciences(Accuracy. 2010) UK: University of Leicester.
2. Bakillah, M., Liang, S. H. L., Zipf, A., & Arsanjani, J. J. (2013). Semantic interoperability of sensor data with volunteered geographic information: A unified model. *ISPRS International Journal of Geo-Information, 2,* 766–796.
3. Bobillo, F., & Straccia, U. (2016). The fuzzy ontology reasoner fuzzyDL. *Knowledge-Based Systems, 95,* 12–34.
4. Bordogna, G., Frigerio, L., Kliment, T., Brivio, P. A., Hossard, L., Manfron, G., et al. (2016). "Contextualized VGI" creation and management to cope with uncertainty and imprecision. *ISPRS International Journal of Geo-Information, 5*(12), 234.
5. Bordogna, G., Kliment, K., Frigerio, L., Stroppiana, D., Brivio, P. A., Crema, A., et al. (2016). Spatial data infrastructure integrating multisource heterogeneous geospatial data and time series: A study case in agriculture, IJGI. *ISPRS International Journal of Geo-Information, 5*(5), 73. https://doi.org/10.3390/ijgi5050073.
6. Bordogna, G., & Pasi, G. (2000). Modeling linguistic qualifiers of uncertainty in a fuzzy database. *International journal of intelligent systems, 15,* 995–1014.
7. Bordogna, G., Carrara P., Criscuolo L., Pepe, M., & Rampini A. (2014). On predicting and improving the quality of Volunteer Geographic Information projects. International Journal of Digital Earth, online edition, 1–22.
8. CCI Cross Community Interoperability- nnex B-CCI Thread scope (2015). Available at: http://www.opengeospatial.org/pub/www/ows10/rfq/annexb-cci.html#cci-thread-scope [10-2-2015].
9. Dubois, A., Raymond, O., Remay, A., & Bendahmane, L. M. (2011). Genomic approach to study floral development genes in Rosa sp. *PLoS ONE, 6*(12), e28455. https://doi.org/10.1371/journal.pone.0028455.
10. Gonzalez, A., Marin, N., Pons, O., & Vila, M. A. (2009). Fuzzy certainty on fuzzy values. *Control and Cybernetics, 38*(2), 311–339.

11. Goodchild, M. F. (2007). Citizens as voluntary sensors: spatial data infrastructure in the world of web 2.0. *International Journal of Spatial Data Infrastructures Research, 2,* 24–32.
12. Gruber, T. R. (1995). Toward principles for the design of ontologies used for knowledge sharing. *International Journal of Human Computer Studies, 43*(5), 907–928. https://doi.org/10.1006/ijhc.1995.1081Citedonpage(s)5,27.
13. Haklay, M. (2012). Citizen science and volunteered geographic information—overview and typology of participation. In D. Z. Sui, S. Elwood, & M. F. Goodchild (Eds.), *Volunteered geographic information, public participation, and crowd sourced production of geographic knowledge*. Berlin: Springer.
14. Hand, E. (2010). Citizen science: People power. *Nature, 466*(7307), 685–687. https://doi.org/10.1038/466685a.PMID20686547.
15. Hitzler, P., Krötzsch, M., Parsia, B., Patel-Schneider, P. F., & Rudolph, S. (2009a). OWL 2 web ontology language primer. W3C recommendation URL http://www.w3.org/TR/owl2-primer/.Citedonpage(s)11.
16. Hitzler, P., Krötzsch, M., & Rudolph, S. (2009b). Foundations of semantic web technologies (pp. 11, 18). Boca Raton: Chapman & Hall/CRC.
17. Klien, E. (2007). A rule-based strategy for the semantic annotation of geodata. *Transactions in GIS, 11,* 437–452.
18. Nativi, S., Craglia, M., & Pearlman, J. (2012). The brokering approach for multidisciplinary interoperability: A position paper. *International Journal of Spatial Data Infrastructures Research, 7,* 1–15.
19. Núñez-Redó, M., Díaz, L., Gil, J., González, D., & Huerta, J. (2011). Discovery and integration of web 2.0 content into geospatial information infrastructures: A use case in wild fire monitoring, in Availability, Reliability and Security for Business, Enterprise and Health Information Systems, LNCS, 6908 (pp. 50–68). Heidelberg: Springer.
20. Ronzhin, S. (July 2015). Semantic enrichment of volunteered geographic information using Linked Data: A use case scenario for disaster management, Master of Science Thesis, University of Twente, Netherlands.
21. Schade, S., & Tsinaraki, C. (2016). Survey report: data management in Citizen Science projects, JRC technical report for European Commission.
22. Stadler, C., Lehmann, J., Höffner, K., & Auer, S. (2012). LinkedGeoData: A core for a web of spatial open data. *Semantic Web, 3,* 333–335.
23. Straccia, U. (2005). Towards a fuzzy description logic for the semantic web. In A. Gomez-Perez, & J. Euzenat (Eds.): ESWC 2005, LNCS 3532 (pp. 167–181) Berlin: Springer.
24. Zadeh, L. A. (1965). Fuzzy Sets. *Information and Control, 8*(3), 338–353.

The manufacturer's authorised representative in the EU is Springer Nature Customer Service Centre GmbH, Europaplatz 3, 69115 Heidelberg, Germany. If you have any concerns regarding our products, please contact ProductSafety@springernature.com

Printed and bound by CPI Group (UK) Ltd, Croydon, CR0 4YY

29/07/2026

02180540-0003